MATERIALS SCIENCE AND TECHNOLOGIES

POST-IMPACT FATIGUE BEHAVIOR OF COMPOSITE LAMINATES: CURRENT AND NOVEL TECHNOLOGIES FOR ENHANCED DAMAGE TOLERANCE

MATERIALS SCIENCE AND TECHNOLOGIES

Additional books in this series can be found on Nova's website under the Series tab.

Additional E-books in this series can be found on Nova's website under the E-book tab.

MATERIALS SCIENCE AND TECHNOLOGIES

POST-IMPACT FATIGUE BEHAVIOR OF COMPOSITE LAMINATES: CURRENT AND NOVEL TECHNOLOGIES FOR ENHANCED DAMAGE TOLERANCE

ALKIS PAIPETIS

AND

DIONYSIOS T. G. KATERELOS

Nova Science Publishers, Inc.

New York

LIBRARY OF CONGRESS CATALOGING-IN-PUBLICATION DATA

Available upon Request
ISBN: 978-1-61668-672-7

Published by Nova Science Publishers, Inc. † New York

Errata for *Post-Impact Fatigue Behavior of Composite Laminate: Current and Novel Technologies for Enhanced Damage Tolerance*

Editor: A. Paipetis and D. T. G. Katerelos;
ISBN: 978-1-61668-672-7

p. v. The Table of Contents should be:

CONTENTS

PREFACE

During their service life, composite laminates are subjected to various transient loadings which often lead to the deterioration of their structural integrity. This deterioration may be manifested in various forms; with increasing impact energy these forms range from delamination to flexural damage to penetration. Among the aforementioned damage modes, delamination is of primary importance as (i) it is invisible and consequently requires non destructive evaluation and (ii) it comprises a considerable damage volume in the affected composite laminate compared to other impact induced damage. The residual properties of laminates deteriorate largely for energies just sufficient to cause penetration; at higher energies, the penetrating impactor passes through the laminate cleanly causing minimum damage away from the penetration area. Furthermore, the impact induced damage is prone to further propagation as the composite is subjected to cyclic loading during its service life. As the propagation of delamination damage is of critical importance for the structural integrity of the composite, it is imperative to predict the delamination growth once present and, at the same time, to develop technologies that minimize the damage initiation and propagation. The scope of this work is to present a thorough overview of the aspects related to post impact damage in composite laminates and its subsequent development under fatigue loading. An introduction on the performance of impacted composite laminates and their constituent phases under cyclic loading will be presented, including special loading cases such as environmental or impact-fatigue. Theoretical semi empirical and numerical models that relate delamination propagation to fatigue loading will be thoroughly reviewed. Then, an overview of the existing technologies that minimize impact damage such as stitching or interleaving will be presented and critically evaluated. Special focus will be given in non destructive technologies to detect damage and its propagation. The state of the art technologies to minimise impact damage will be presented together with the novel concept of hybrid composite systems that employ nano scale phases in order to enhance specific energy absorption and minimize damage propagation. Finally, the applications of damage tolerance technologies in composite structures will be discussed.

ABBREVIATIONS

AE	Acoustic Emission
CAI	Compression After Impact
CDM	Continuum Damage Mechanics
CF	Carbon Fibre
CFRP	Carbon Fibre Reinforced Plastic
FBG	Fibre Bragg Grating
FEA	Finite Elements Analysis
GF	Glass Fibre
GFRP	Glass Fibre Reinforced Plastic
HVI	High Velocity Impact
LVI	Low Velocity Impact
NDI	Non-Destructive Inspection
OM	Optical Microscopy
SEM	Scanning Electron Microscopy
SERR	Strain Energy Release Rate

Chapter 1

INTRODUCTION

Impact as a cause of premature failure of air structures is of major importance. One of the most important air tragedies, the Concord accident that occurred in Paris in 2000 was attributed to impact damage induced by a tire fragment which hit the jet's fuel tanks. Composite materials, and particularly polymer matrix laminates, are particularly vulnerable to foreign object damage; in many cases, impact results in invisible damage due to the inherent through thickness anisotropy of the laminated structure in conjunction with the brittleness of the matrix. In this case, impact may result to a series of failure modes, not all of which are straightforwardly detected. Low-velocity impact (LVI) in particular is characterized by non catastrophic failure of a specimen; specimens subjected to LVI do not suffer penetration and still exhibit a considerable amount of residual strength [1]. This is expected since LVI involves by definition relatively low impact energy. However, LVI is well known as a major cause of delamination onset. When a projectile or an impactor hits a laminate with low velocity, the localized contact stresses lead to large local out-of-plane stress components. At higher velocities, this phenomenon is driven by localized contact stresses and the propagation of a generated internal stress wave. This stress wave is the cause of delamination initiation at the ply interfaces, where there is a major change of ply stiffness due to the change in the ply orientation [2, 3]. The amount and the type of damage in the laminate depend upon the mass and geometry of the impactor [4, 5] as well as the configuration of the laminate [2].

In general, composite damage is categorized into four main groups, matrix cracking, interlaminar failure (delamination), interfacial failure (fibre-matrix debonding) and fibre failure. While broken fibres are seldom observed after low or moderate velocity impact, matrix damage such as intralaminar cracks and delaminations are always present bellow the apparently intact surface layer of a composite structure (Figure 1). Both these damage modes could be catastrophic for the structure at a secondary level. During the service loading of the structure, local stress concentrations and strain magnifications are created at the vicinity if the impact induced defects, which in their turn give rise to secondary damage modes. These damage modes lead to the reduction of the load bearing and the load transfer ability of the structure and potentially include fibre failures [6], which accumulate to lead to premature failure.

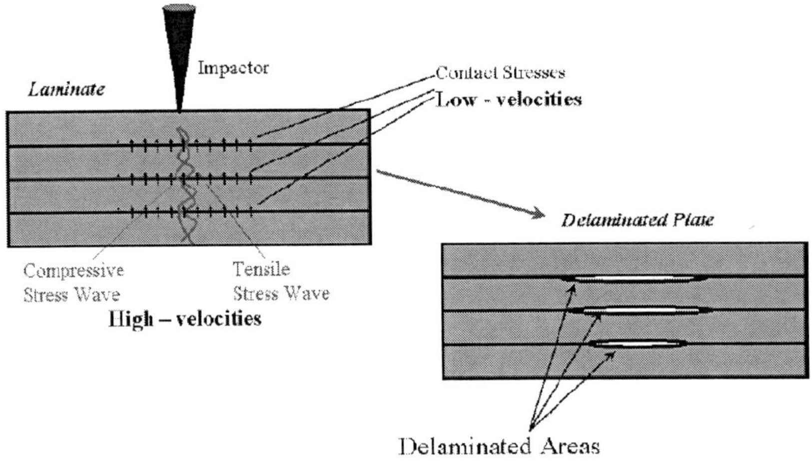

Figure 1. Blind damage induced by Low Velocity Impact (LVI).

Delamination between the constituent laminae of a composite laminated structure is categorized into two groups, interior and exterior. Interior delamination appears far from the structure edges and is mainly caused by incomplete curing, the introduction of foreign particles during manufacturing, or matrix cracks in the longitudinal and the transverse directions that propagate during operation and bridge each other forming a delaminated area. Exterior delamination is due to edge effects, i.e. stress singularities of the out-of-plane stresses developing at the edges of a laminate in the arising stress fields. The 'interlaminar stress field' plays a dominant role as far as delamination onset and growth is concerned. This field refers to the stresses arising through the thickness of the laminate and transverse to the plane of the laminate. The onset and growth of delamination after impact is also of major importance as a damage tolerance design parameter.

The aim of this study is to review the up-to-date work and present the recent developments in relation to the post-impact damage initiation and growth in polymeric composite laminates. More analytically, the following fields are considered in detail. A review of the impact response of polymer matrix materials is performed, as matrix properties are of primary importance to the performance of the composite,. The impact behaviour of composite laminates is then examined, taking into consideration thermal effects, natural composites and impact-fatigue. Theoretical models for the delamination growth under fatigue loading are presented. Special emphasis is given to the inspection techniques for delamination – and damage in general – developed within a structure, i.e. methodologies for the measurement, characterisation and growth quantification. To this end, a detailed investigation of the non-destructive inspection (NDI) methods, such as ultrasonics, acoustography, electrical measurements, optical and imaging techniques is performed. Works that describe the post impact behaviour under static and cyclic loading are summarized and available delamination propagation models are reviewed. A special mention is made for the case of natural composites. An overview of the existing technologies that minimize impact damage, i.e. z-pinning, stitching, interleaving and matrix toughening is presented and critically evaluated. The concept of hybrid composite systems that employ nano scale phases in order to enhance specific energy absorption and minimize damage propagation is presented and elaborated. Finally applications of the post-impact/ fatigue behaviour of specific systems, such as composite joints and innovative materials, like GLARE, are detailed.

Summarising, this work presents an integrated approach on the impact and post impact behaviour of composite structures, starting from the causes of the impact damage initiation and its propagation due to post impact thermo-mechanical loading, continuing with the damage identification and quantifying procedures, elaborating on the available studying and predicting methodologies and closing by critically evaluating the available remedying and preventing technologies.

Chapter 2

IMPACT RESPONSE

A. POLYMERS

The polymeric matrix of a laminate is of primary importance regarding the impact response of the composite structure. During the last decade, a series of works have been published examining and explaining the parameters involved in the impact response of polymeric materials used as matrices in composite materials.

Perkins [7] reviewed the factors that influence the impact resistance of polymers, emphasizing on crystalline polymers. The phenomenology of polymeric fracture was examined, including brittle and ductile failure and the ductile-brittle transition temperature (DBTT). An in-depth discussion of the effects of crystalline morphology was presented, with special attention to the influence of spherulite size, fillers, processing conditions, transitions/relaxations, and multilayer co-extrusion. Rubber phase addition was considered, including mechanisms, morphology, rubber type and particle size, and test conditions. Finally, common impact test methods were surveyed, including pendulum, falling weight, tensile impact, and tensile elongation tests. General guidelines were listed in the review conclusions for improving the impact resistance of a polymeric system: (i) microscopical examination of the fracture surfaces to ascertain the fracture mechanisms under conditions similar to its intended use, (ii) determination of the impact test correlated with the type of fracture, (iii) reviewing the processing procedures and determination whether the morphology imparted during fabrication can be improved, (iv) investigation the addition of various types of fillers, (v) determination if manufacturing procedures benefit or harm the structure, (vi) consideration of deformation mechanisms that may provide enhanced ductility and impact resistance.

Arogn and Cohen [8] demonstrated that all solid polymers are intrinsically brittle and undergo a ductile to brittle fracture transition based on the nature of their bonding alone. The most effective way of avoiding a ductile to brittle transition is to reduce the plastic resistance to delay reaching the brittle strength which in unoriented polymers is governed by intrinsic cavitation. While a number of possibilities for this exist, the most widely used techniques involve incorporation of rubbery particles or rigid particles. In both approaches the continuous homo-polymer is transformed into a quasi-regular cellular solid that is much more capable of undergoing large local plastic flow by ligament stretching between cavitated particles and is less susceptible to the propagation of brittle cracks under the usual conditions of tensile straining. In many instances, as in epoxies with high flow stresses and strain hardening rates, such cavitation alone can result in attractive levels of toughening by crack tip shielding. In semi-crystalline polymers, however, dramatic effects can be achieved with either rubbery or rigid particles when the quasi-uniformly cavitated cellular polymer can undergo wide spread large strain plastic extension of inter-particle ligaments in slow tension resulting in impressive crack extension resistances J. Arogn and Cohen [8] proved that, under impact conditions, a notched specimen of rubber particle-modified polymer can still exhibit considerable toughening. However rigid-particle-modified polymers suffered severely from clustering of rigid particles into super critical flaws that trigger brittle response, much like what is encountered in structural steels. Based on their known mechanical response in neat form, six semi-crystalline polymers were analyzed in detail to evaluate their potential for toughening under impact conditions. The results correlated very well with experimental findings.

B. Fibrous Composites

An experimental investigation was presented by Chotard and Benzeggagh [9]. They studied the dynamic behaviour of composite pultruded glass/polyester profiles subjected to low-velocity impact (LVI) with several incident kinetic energy ranges. The results of the drop-weight tests provided specific information about the effects of the impactor velocity, mass and nose radius on the impact response of the profiles. A big-impactor (BI) strike produced less delamination than a small-impactor (SI) one. However, the location of delaminated areas was different for BI and SI impacts. Thus, damage localisation had more influence on the mechanical response of these profiles than the damage extent itself. An experimental study of the sequence of damage mechanisms during impact loading was also carried out. The damage sequence over the impact duration was identified using a methodology based on combined accelerometer/strain-gauge signal analysis. This analysis allowed for the explanation of the impact event and the identification of damage chronology. The accelerometer provided displacements values used to determine the vibration energy. The attenuation (loss) factor calculated from the acceleration signal and its variation was correlated with damage extension. The chronology of separated damage mechanisms was determined with good accuracy. Two different types of failure -cracking and delamination- were observed. Finally, an inverse variation of damage characteristics versus velocity and energy magnitude was emphasised, which demonstrated the separate effects observed between mass and velocity characteristics and the influence of experimental parameters on the impact response of pultruded beams.

The effect of laminate structural geometry on impact performance of woven aramid fibre - polyethylene (PE) fibre/vinylester hybrid composites was examined by Park and Jang in [10]. The impact properties of one- and two-layer hybrid composites were compared with respect to impact mechanism and deformation extent. In one-layer composites, two intraply hybrids exhibited different deformation extent and load transfer at crossover points. Intraply fabric hybrid laminates with the same amount of aramid and PE fibres absorbed most of impact energy through large deformation of PE fibres with an elliptical damage shape. The bundle of PE fibres existed around the perforated point with a slippage of PE fibres toward the impact point. The laminate with a 3:1 ratio in the aramid/PE fibres volume fraction showed the higher impact energy because both aramid and PE fibres contributed to the absorption of impact energy with a round damage zone and the load transfer to the PE fibre direction was restricted by adjacent aramid fibres. The laminate exhibited aramid fibre breakages near the impact point with little slippage of PE fibres. In the case of two-layer composites, interply hybrid composites exhibited higher impact energy than intraply hybrid composites. The impact energy of interply hybrid composites was mainly absorbed through delamination, and thus impact energy was well correlated to delamination area. The impact energy of intraply hybrid composites was primarily dominated by full exertion of deformation in PE fibre rather than a delamination process.

Dutta et al [11] observed the fracture morphology and texture of graphite/epoxy composite fragments after impact. They identified and explained the observed changes in the surface texture of the fragments for different impact velocities (122 to 610 m/s, 400 to 2000 ft/s) and over a wide range of specimen temperatures (-54 to 24 °C, -64 to 75 °F). The composite panels were impacted by spherical steel projectiles. The spall was differentiated according to different sizes and shapes. A few fragments representing each shape and size were selected to analyze the surface morphology using SEM. Changes in the surface texture were observed according to the different sizes and shapes, and the change in size and shape of the fragments was credited to the change in impact force. Unidirectional panels absorbed more energy at low velocity range. A series of interesting conclusions were drawn from the experimental data. Cross-ply laminates absorb more energy than 6061-T6 aluminium at room temperature, but not at low temperature. At low temperature, composites require much less energy to delaminate per unit delaminated area. The energy absorbed by quasi-isotropic panels was higher than the energy absorbed by cross-ply panels indicating that toughness increases with multiple orientations of the fibres. Of the three different types of surface textures, i.e., gravel, hackles and matrix rich areas, gravel was the dominant morphology. At low temperature, because of clean debonding, the smooth matrix area was more common; while at room temperature the matrix-rich area was relatively rough (Figure 2).

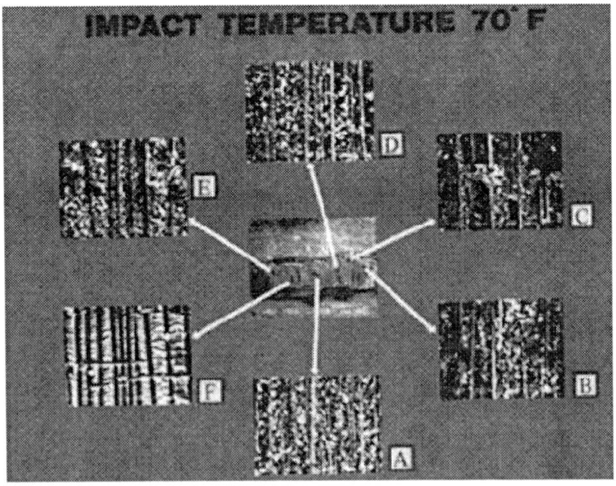

Figure 2. Texture changes with depth for impacted CFRPs at room temperature [11].

The fracture behaviour and failure mechanisms of composite laminates containing woven fabrics in mode I and mode II delamination and under impact loading were investigated in the overview performed by Kim and Sham [12]. Potential advantages of using woven-fabric composites over those made from angle-ply unidirectional prepreg tapes were discussed in terms of the interlaminar fracture toughness, the impact damage resistance and damage tolerance (Figure 3). Angle-ply laminates possessed higher mode I interlaminar fracture toughness, G_{IC}, than (0°/0°) unidirectional laminates as a result of the extra toughening mechanisms associated with crack branching into the adjacent layers, leading to multiple delamination. Woven-fabric laminates also showed much higher G_{IC} values (often more than 4 – 5 times) than their unidirectional counterparts. Unique features and advantageous failure mechanisms were identified: inherent roughness of the fabric; the availability of resin-rich regions between the fabrics; crack propagation along the undulating pattern of the yarns creating a large fracture surface area; and multiple crack front formation. The mode I interlaminar fracture behaviour of glass-woven-fabric composites was shown to be controlled by the fibre-surface treatment. A weak interfacial bond offered by a low silane concentration promoted the stability of crack propagation, with high interlaminar fracture toughness. The major failure mechanisms taking place in mode II delamination were fibre/matrix interface debonding in shear, and fracture of matrix resin between the reinforcing fibres. There was little dependence of G_{IIC} value on interply angle as this was dominated by the interfacial bond quality. The glass-woven-fabric laminates containing high silane concentrations had a strong interfacial bond with high G_{IIC} values. Woven-fabric laminates in general exhibited a lower maximum load, a smaller damage area, a higher ductility index, and a higher residual compression-after-impact (CAI) strength than cross-ply laminates. Higher mode II interlaminar fracture toughness, lower thermal and elastic mismatches along with a more ductile and compliant nature of the fabrics were mainly responsible for these observations for woven-fabric laminates. The effects of fibre/matrix interface properties influenced by different silane coupling agents on interlaminar fracture and the impact performance of woven-glass-fabric composites were evaluated. A higher silane concentration results in improved impact performance of the composite through its ameliorating influence on G_{IIC} values.

A plane stress continuum damage mechanics (CDM) model for composite materials was implemented in the non-linear finite element analysis (FEA) code, LS-DYNA3D and its predictive capability was evaluated by carrying out a series of numerical simulations involving impact of laminated composite plates [13]. The performance of the model relative to that of an existing model in LS-DYNA3D was assessed by comparing the computed results with instrumented impact test results. It was demonstrated that a CDM based approach offered a versatile tool for predicting damage progression in composite structures. Although the parameters used in such models are often non-physical and difficult to characterize, the formulation presented a unique and interesting approach to the numerical simulation of damage in composite materials. According to the authors [13], the work reported represented the first step in an ongoing effort to develop a versatile, numerically robust, and physically meaningful CDM based constitutive model for composite materials.

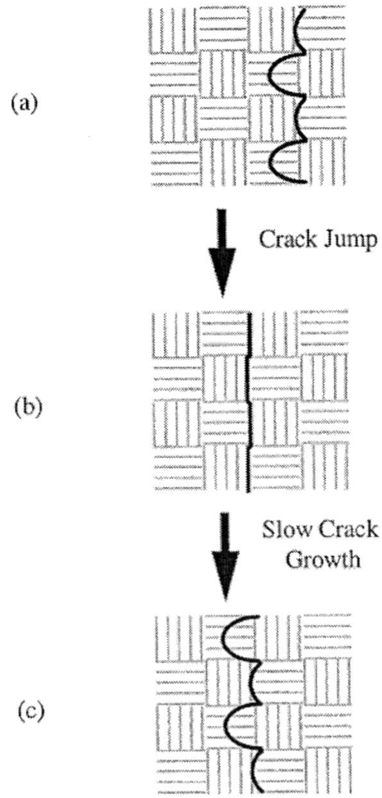

Figure 3. Schematic illustration of the sequence of delamination crack propagation in a woven fabric laminate [12].

Morozov et al [14] studied the damage resistance of laminated composite helicopter blades subjected to impact loading using combined theoretical and experimental approaches. Dynamic stress intensity factors were determined for composite laminates and separate layers as functions of the size of the defects caused by the projectile impacts. The effect of the projectile size on the damage tolerance of composite blades was investigated and tolerable sizes of the defects were estimated. The methodology adopted for the study proved capable to produce results applicable to the analysis and damage tolerant design of composite structural components subjected to impact loading by projectiles. The comparison of the experimentally determined critical values of stress intensity factors with the computational results yielded maximum allowable sizes of the defects depending on the size of the projectiles.

Corum et al [15] attempted to experimentally characterize the susceptibility of three candidate automotive structural composites to incidental low-energy impact damage. The composites, each of which had the same urethane matrix, were produced by a rapid moulding process suitable for high-volume automotive applications. The reinforcement for the first composite was a random chopped-GF, while the remaining two were reinforced with stitch-bonded carbon-fibre (CF) mats, one in a cross-ply layup, and the other in a quasi-isotropic layup. A pendulum device, representative of events such as tool drops (LVI), and a gas-gun projectile, representative of events such as kick-ups of roadway debris (high velocity impact – HVI), were used to impact plate specimens. In both cases, the impactor point was a 12.7-mm-diameter hardened-steel hemisphere. The effect of hydrothermal exposure on the impact performance of the specimens was also investigated. Brick-drop tests were performed to assess the applicability of the baseline pendulum and gas-gun data to other events. Following the impacts, the damage areas were measured and the plates were cut into either tensile, standard compressive, or CAI specimens for determining strength degradation. The GF composite was least susceptible to damage, followed by the cross-ply CF laminate with the same thickness. The quasi-isotropic CF composite, which was thinner than the other two, exhibited the maximum damage. While compressive strength was significantly degraded by moderate damage in the random-GF composite, tensile strength was not, since fibre breakage rarely occurred in the damage region. On the other hand, both tensile and compressive strengths were degraded in the cross-ply CF laminate. The compressive strength degradation for a given damage area was similar in the two CF laminates. Both exhibited less degradation than the GF composite. For the quasi-isotropic CF laminate, it was shown that strength degradation produced by an open circular notch provided a reasonable lower bound to the degradation due to an impact damage area of the same size. This was in line with the common assumption often used in damage tolerance evaluations [16], that impact-induced defects introduce the same strength reduction as a hole of the same size (Figure 4). While damage in the moulded random GF composite correlated well with kinetic energy, the same was not true for the CF laminates. For these composites, the damage area produced by the pendulum was about twice that produced by the gas-gun projectile for the same kinetic energy. To correlate the aforementioned experimental findings, the parameter $m^a v$ was employed, where m and v were the impactor mass and velocity, and the exponent a was a constant. A single baseline curve relating damage area to impactor mass and velocity was obtained for each composite. These curves were also shown to provide adequate approximations of the damage produced in pre-soaked specimens and specimens tested at -40 °C. With the significant exception of two tests, they gave generally conservative approximations for damage produced by brick drops.

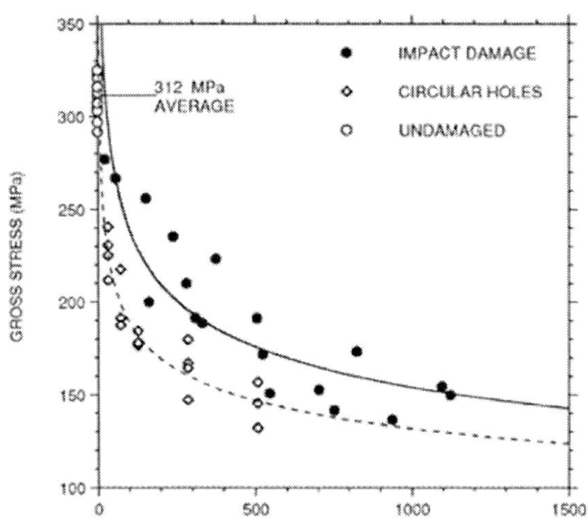

Figure 4. Comparison of CAI strength degradation due to circular holes in quasi isotropic laminates to that due to impact damage [15].

The suitability of damage – force maps for characterising LVI damage in a glass/polyester composite was investigated at low and intermediate energies [17]. Tests were performed on circular plates with diameters ranging between 50 and 300 mm and on square plates with edge lengths of 75 and 200 mm.

Damage appeared in the form of delamination under the point of impact and widespread matrix cracking. The translucent nature of the composites facilitated the determination and quantification of the extent of damage within each specimen. Plots of delaminated area against impact force yielded maps in which the experimental data laid within a narrow band over the range of conditions examined. For a given impact energy, the impact force was greater and damage was more severe in smaller coupons. The damage – force maps showed that the impact force to generate delamination lay between 600 and 800 N. An energy-balance model was used to successfully predict the impact response of circular structures and to predict the onset of damage within these composite plates. The model could predict the impact response of targets that incur reasonably high levels of damage and provide an initial estimate of the impact conditions required to initiate interlaminar failure in fibre-reinforced composites.

Kang et al [18] dealt with the effect of constituent materials on impact damage and the strength reduction of a sandwich structure composed of laminated face sheets and Nomex® honeycomb core. LVI tests were performed and damage was inspected by scanning acoustic microscopy. In addition, static tests were conducted under flexural loading. The damage behaviour was characterized through the energy absorbed during impact. The strength reduction was evaluated via a residual strength prediction model and was found to depend both on the face sheet material and core thickness. A statistical model was developed to identify the variation of residual strength of the impacted structure. It was shown that the residual strength was affected by both the face sheet material and core thickness.

Quasi-isotropic and cross-ply glass/epoxy laminate plates were investigated in crush and LVI crush tests [19].The characterization of the damage was performed in relation to the type of test, the ply stacking sequence, the plate dimensions and the maximum force reached during impact. There was a strong correlation between the force-displacement curves from the crush tests and impact tests. The maximum load caused the identical deflection in both tests corresponding to the same maximum force, therefore indicating that the deformation was not sensitive to the velocity of the load application. For a specific laminate sequence, damage parameters were similar for crush and impact tests when flexibility was less than a threshold value. From this and the previous statement, it was concluded that for laminate structural components with certain boundary conditions and whose dimensions and thickness give a flexibility factor under a certain value, crush tests are a viable alternative for evaluating low velocity object impacts. In the case of highly flexible components, although force displacements curves were similar, damage area and length were different for the same lamination sequence and so the aforementioned conclusion was not valid. The quasi-isotropic plates in general supported higher maximum loads than cross-ply specimens, although, in general, they exhibited greater damage. The increased non-linear force/deflection behaviour of the most flexible plates was verified. For a Delamination Threshold Load, the duration of the impact depended on the transverse flexibility of the plate. Contact time increased with flexibility. The shape of the damage depended on the fibre orientation, and could be approximated with a circle for low flexibility. The through thickness distribution of damage for the stiffest plates was of the typical trapezoidal shape with the longest base in the face opposite to the load application. For the most flexible plates, damage was practically constant across the thickness. The indentation depth increased with decreasing flexibility of the plate. Finally, plates absorbed more energy in the crush tests than in the impact tests and that difference was more prominent with decreasing flexibility.

Effect of Temperature on Impact Response

The effect of temperature variations for both low and high temperature ranges was studied experimentally in relation to impact damage to carbon fibre reinforced plastic (CFRP) laminates [20]. CF/epoxy orthotropic laminated plates with cross-ply lay-ups ($[0_6/90_6]_s$ and $[0_4/90_4])_s$, and CF/PEEK orthotropic cross-ply laminated plates ($[0_4/90_4]_s$) were used. A steel ball launched by an air gun inflicted the impact damage on the CFRP laminate. Non-destructive evaluation (NDE), such as scanning acoustic microscopy (SAM) was performed on the delamination-damaged samples to characterize damage growth at different temperatures. The increase in the temperature of a CFRP laminate resulted in decreasing impact induced delamination areas.

Gómez-del Rio et al [21] examined the impact response of CFRP laminates at LVI under low temperature conditions. Square specimens of CF/epoxy laminates with different lay-ups (unidirectional, cross-ply, quasi-isotropic and woven laminates) were tested using a drop weight device. The test temperature ranged from 20 down to –150 °C. After the impact tests, the damage extension was measured by C-Scan and the damage mechanisms were studied by optical and scanning electron microscopy (OM and SEM, respectively). The influence of temperature, ply reinforcement architecture and stacking sequence on the mechanical behaviour of the CFRP laminates subjected to low velocity impulsive loads was evaluated (Figure 5). The experimental results obtained showed that cooling the laminate before impact had an effect on damage similar to that of increasing the impact energy, i.e., larger matrix cracking and delamination extension, deeper indentation on the impacted side, and more severe fibre – matrix debonding and fibre fracture on the opposite side. However, an essential difference between both damage-increasing effects was observed; whereas the energy absorbed by the laminate was roughly proportional to the impact energy, larger damage extent favoured by low temperatures was not followed by a greater dissipation of energy due to the lower specific fracture energy of the epoxy matrix at cryogenic conditions. The threshold energy decreased up to 50% in quasi-isotropic laminate when temperature fell from 20 to -150 °C. This thermally induced effect was severe in tape laminates (cross-ply and quasi-isotropic), where orientation of the carbon fibres was different in each layer. In these materials, low temperatures produced interlaminar residual thermal stresses, high enough to accelerate matrix cracking and delamination during LVI. For multidirectional laminates manufactured with plain woven unidirectional fabric, the aforementioned phenomenon was not as prominent, and damage extension after impact did not increase substantially at low temperature.

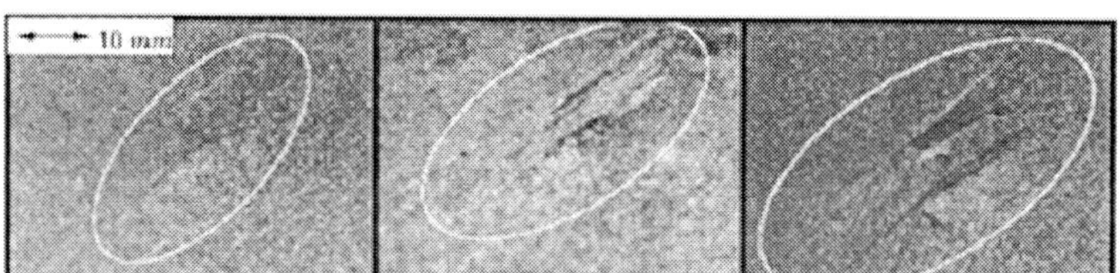

Figure 5. Effect of temperature on impacted quasi isotropic plates 20 °C (left), -60 °C (center) and -150 °C (right) [21].

Natural Composites

Sabeel Ahmed et al [22] explored experimentally the effects of hybridization of GF on LVI behaviour and damage tolerance capability of woven jute fabric reinforced isophthalic polyester composite. Laminates were fabricated by hand lay–up and cured under light pressure at room temperature. LVI tests were conducted on jute and hybrid samples (150×150mm) using an instrumented drop weight impact tower. Some of the samples were subjected to the NDI (C-scan) to study the nature and extent of damage and to measure the delamination area. Post-impact tension tests were conducted to assess the damage tolerance capability of the composites. The results of the study indicated that jute laminates had better impact energy absorption capacity than jute–glass hybrid laminates; however their damage tolerance capability was less than that of jute–glass hybrid laminates.

Impact-Fatigue

Impact-fatigue as a distinct case of fatigue is of major importance, since it can simulate realistic working conditions of composite components. As it is correlated to fatigue [23], it will occupy a part in the present review. The theoretical analysis of impact-fatigue phenomena was presented [24] and the resultant fatigue equation was tested against experimental impact-fatigue data. Fatigue life proved to be a function of fibre orientation with respect to the impact load direction. There was a minimum in impact energy for failure to occur, or an impact-fatigue limit. The lifetime equation of impact-fatigue was derived and analysed statistically to predict the engineering lifetime for design purposes. As postulated, impact-fatigue behaviour can be analysed by means of fracture mechanics concepts, because crack propagation is the governing mechanism [25].

One of the very first studies of impact-fatigue was performed on 63.5% GF/vinylester resin, notched composites [26]. A well-defined impact-fatigue (S–N) behaviour, having a progressive endurance below the threshold single cycle impact fracture strength and possessing a fatigue limit was experimentally defined. Fractographic analysis revealed fracture by primary debonding, with fibre breakage and pull-out in the tensile zone and a shear fracture of fibre bundles in the compressive zone of the specimen (Figure 6). The residual strength measured after impact-fatigue showed initially retention of the strength at high impact energy levels and then a gradual drop followed by a rapid drop. The residual modulus and toughness showed a gradual drop with increasing number of impacts. It was suggested that few large cracks for high impact endurance and an increased volume of microcracks in the matrix for low impact endurance account for failure of the composite under impact-fatigue.

Azouaoui et al carried out an investigation [27] on the impact-fatigue damage of glass/epoxy laminated composites. Damage accumulation, such as matrix cracking, delamination and fibre breakage was the immediate result of repeated impact of the composite material and led to the reduction of the overall stiffness. A model was proposed for characterizing damage as a function of the number of impact events. The macroscopic failure mode and the internal damage in glass/epoxy laminated specimens as a consequence of impact-fatigue were analysed for different levels of incident impact energy.

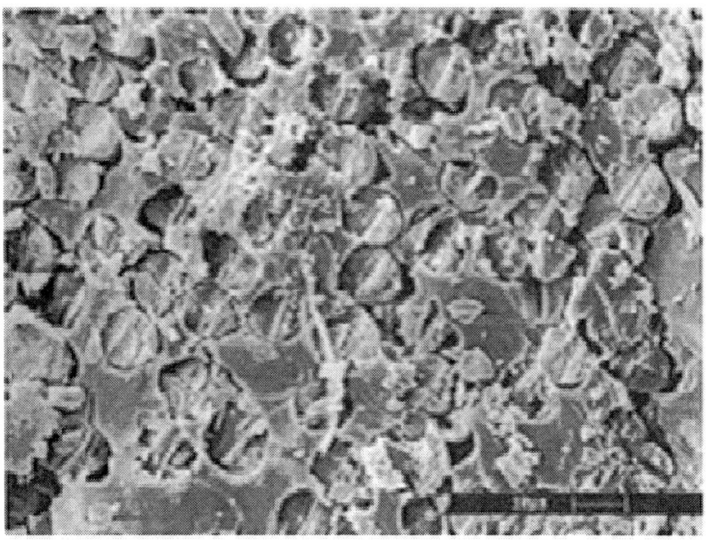

Figure 6. SEM micrographs of composite plates submitted to impact-fatigue showing shear failure of the fibres [26].

In the work presented by Barkoula et al [28], the influence of stacking sequence, existence and position of interleaves on the solid particle erosion in CFRP composites was investigated. The erosive wear behaviour was studied in a modified sandblasting apparatus at 90° impact angle. The erosion behaviour was considered as a repeated impact process or impact-fatigue. A semi-empirical approach initially developed for the prediction of the residual strength after single impact was adopted and evaluated in the case of erosion conditions. The excellent agreement between theoretical predictions and experimental values proved the reliability of this model as a useful tool for the prediction of the post impact residual strength in the case of solid particle erosion (Figure 7).

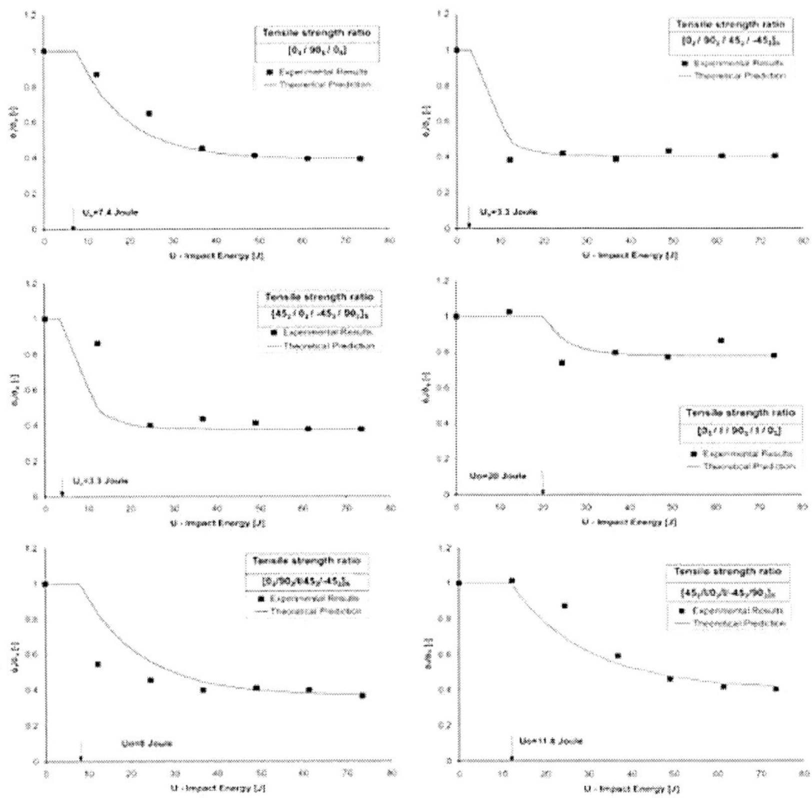

Figure 7. Normalised residual tensile strength versus impact energy and model predictions: various laminate configurations [28].

Schrauwen and Peijs [29] described the results of falling weight impact tests on GF-reinforced laminates. The test program consisted (i) of falling weight impact tests for the determination of the penetration energy and the influence of laminate construction on damage development and (ii) repeated falling weight impact tests for the determination of the impact-fatigue life and damage development under repeated impact conditions at sub-penetration energy levels. The objective of the work was to compare the impact behaviour of cross-ply laminates based on a brittle unsaturated polyester resin and a more ductile vinylester resin system with two types of glass reinforcement, i.e. woven and multiaxial non-crimp fabric. The penetration energy of the various composite laminates depended on the type of reinforcement, whereas damage development during repeated impact was strongly influenced by both fibre architecture and resin. No significant effect of the different material parameters investigated on the number of impacts to penetration was observed. Especially when the repeated impact energy was normalised with respect to the penetration energy, all laminates exhibited similar behaviour.

Williams et al [30] focussed on the development, implementation, and verification of a plane-stress continuum damage mechanics (CDM) model for composite materials. A physical treatment of damage growth based on the extensive experimental literature on the subject was combined with the mathematical rigour of a CDM description to form the CDM model. The model was implemented in the commercial FEA code LS-DYNA. Its prediction of impact damage growth and its representation on the impact force histories in carbon fibre reinforced plastic laminates were shown to be physically meaningful and accurate. Furthermore, it was emphasised that the material characterization parameters could be extracted from the results of standard test methodologies, for which a large body of published data is available in the literature.

The impact-fatigue properties of three kinds of GF or glass-bead reinforced polymers were studied [23] via uniaxial and cyclic indentation tests with concurrent Acoustic Emission (AE) and visco-elastic property measurements. The detailed failure mechanisms and damage development mechanisms were investigated with acoustic velocity measurements and OM observations. Prior to fatigue tests, the materials were characterized by basic mechanical tests such as tensile tests, indentation tests, and drop weight tests. When performing fatigue tests, special attention was paid to the effect of interval times between loading on the fatigue properties of the composites. It was found that the numbers of cycles to failure were strongly dependent on the duration of the interval time. In the case of uniaxial fatigue tests, longer interval times resulted in larger numbers of cycles to failure. On the other hand, in cyclic indentation tests, longer interval time resulted in smaller numbers of cycles to crack initiation and ultimate failure. The results of AE measurements were in good agreement with the results of cycles to failure in fatigue tests and the test conditions. Attention was also paid to the relationship between the size of the damage zones and interval time between loadings. The difference in the cycles to failure due to the different loading modes and interval time were understood by the differences in the mechanisms of damage zone development, where breakage of glass, micro-voiding, change in the orientation of glass fibres, and plastic deformation of matrix occurred. The degree of damage localization had a strong effect on the fatigue life. The differences in the damage development mechanisms were attributed to the difference in the elastic response of the specimens due to the different loading mode (uniaxial or indentation) and the interval time.

Repeated LVI drop weight tests were carried out [31] on advanced composites (glass, carbon, Kevlar® in epoxy matrix) using an instrumented impact test machine. The number of drops to failure (N_f) data was obtained with simultaneous recording of the load–time and energy–time traces. The extent of impact damage for glass and Kevlar® composites was interpreted using final delamination area maps or tracings. Results for all three composites showed that with an increase in drop numbers, the peak load (P_L) steadily decreased while the total energy (E_t) to failure increased. Also, as the incident energy (E_{in}) was varied in arithmetic progression, the number of drops to failure (N_f) varied in harmonic progression. It was concluded that repeated drop tests with final delamination area maps assist in understanding the impact damage tolerance of polymer composites.

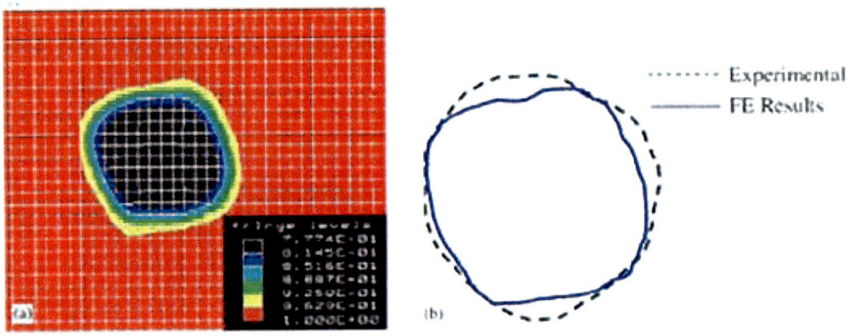

Figure 8. Impact damage area (a) matrix cracking mode (0 failed-1 elastic) and experimental versus FE results [32].

Meo et al [32] described the results of an experimental investigation and a numerical simulation on the impact damage for a range of sandwich panels. The test panels were representative of the composite sandwich structure of the engine nacelle Fan Cowl Doors of a large commercial aircraft. The LVI response of the sandwich panels was studied at five energy levels, in order to assess damage initiation, damage propagation, and failure mechanisms. A numerical simulation was performed using LS-DYNA3D transient dynamic FEA code for calculating contact forces during impact along with a failure analysis for predicting the threshold of impact damage and initiation of delamination. Good agreement was obtained between numerical and experimental results (Figure 8). In particular, the numerical simulation was able to predict the extent of impact damage and impact energy absorbed by the structure. It was shown that a rigid numerical model can yield significant information towards the understanding of the mechanisms involved in LVI even prior to testing, and therefore aid to designing a more efficient impact-resistant aircraft structure.

The effect of repeated LVI on the performance of carbon/epoxy composites with three different stacking sequences was evaluated [33]. The evolution of the macroscopic damage was characterized by a phenomenological quadratic equation. The coefficients of the equation were parametrically adjusted to the experimental points and related to the degree of freedom of the composite to deform, the rate of change of the composite properties due to the impact events, and the change of the damage mode. These coefficients were shown to be dependent on the laminate stacking sequence and successfully described the macroscopic behaviour of the composites. The experimental results showed that the cross-ply and non-symmetric laminates had better endurance against LVI than unidirectional laminates. As was expected, this result was due to multidirectional reinforcement which does not allow for "easy paths" for cracks at matrix rich areas. As for unidirectional laminates, fast trough thickness cracks developed, rapidly leading to failure. For all studied configurations, the surface crack pattern followed the fibre orientation of the top laminae, indicating matrix and fibre-matrix interface rupture. Moreover, the cross-ply non-symmetric laminates developed huge surface indentations, easily observed by naked eye, well in advance of their failure. Therefore, these two configurations were safer with respect to the repeated LVI.

The influence of laminate thickness on the resistance to repeated low energy impacts of glass, carbon and aramid fabric reinforced composites was evaluated for two levels of impact energy [34]. According to the study, below a certain energy level the cross section of the laminate was the most dominant parametre that determined impact resistance. For the same cross section, the experimental points for all tested laminates fell on a single curve, irrespective of the reinforcement. When the energy level of the impactor was increased, fibre characteristics became significant. The glass fabric reinforced composite showed the steepest increase on the impact resistance with increasing laminate thickness. This behaviour was attributed primarily to the higher areal coverage of the glass fabric used. The more isotropic behaviour of GF in relation to the anisotropic character of aramid and CF was also found to be of relevance. The experimental data were fitted against laminate thickness using a quadratic equation [33].

Studies were undertaken on the effects of different post-cure schedules on the impact resistance of room temperature-cured glass/epoxy laminates subjected to repeated drop tests, using an in-house built un-instrumented drop weight impact tester [35]. The impact resistance was assessed in terms of two parameters, viz., the number of drops to failure (N_f) and the delamination area growth (dA). The specimens were post-cured at different time–temperature combinations and the impact response was studied at three different incident energy levels. It was found that, the N_f and dA values increased to a maximum at an optimum post-cure schedule of 85 °C/4h. The extent of composite cure for the different post-cure schedules was ascertained by measuring the glass transition temperatures (T_g), in order to obtain a correlation between the impact resistance and the chemical cure status of the composite. In addition, single drop impact studies were made using an instrumented impact tester and the efficiency of the optimum post-cure schedule was confirmed.

The objective of the experimental campaign presented by Baucom et al [36, 37, 38] was to obtain a detailed understanding of the effects of reinforcement geometry on damage progression in woven composite panels under repeated drop-weight impact loading conditions. The composite systems included a 2D plain-woven laminate, a 3D orthogonally woven monolith, and a biaxially reinforced warp-knit. The radial spread of damage was smallest for the 2D laminates and largest for the 3D woven composites. The 3D composites had the greatest resistance to penetration and dissipated more energy in total than the other systems. This was attributed to unique energy absorption mechanisms, which involved the crimped portion of z-tows in the 3D composites implying that failure was controlled by the configuration of the z-tows. The 3D systems exhibited both an inherent capability to dissipate energy over a larger area and greater perforation strength than other systems with comparable areal densities and fibre-volume-fractions.

Belingardi et al [39] investigated the response of a non-symmetric GFRP to repeated impacts. The laminate was intended for nautical applications, was manufactured using hand lay-up with subsequent vacuum bagging and possessed different matrix systems on each side, i.e. polyester and vinylester. With the aim of looking into the role of the matrix, two series of static indentation and repeated impact tests were performed, impacting the vinylester face and the polyester face of the laminate respectively. For the impact tests, four impact velocities were considered, and a minimum of four specimens for a given velocity were subjected to forty repeated impacts or up to perforation. The impact response was evaluated in terms of damage progression by visual observation of the impacted specimens, evolution of the peak force and of the bending stiffness with the number of impacts and by calculating the damage degree, defined as the ratio between the absorbed energy and the impact energy. For impact velocities for which no perforation occurred within the test duration, the experimental data for the two series essentially overlapped. On the contrary, for perforation tests, the vinylester resin exhibited higher peak forces but a lesser number of impacts to perforation. Visual observation of the impacted specimens revealed that, in both series of tests, lamina splitting occurred at the interface between the two resin systems. Moreover, the delamination area grew very rapidly during the first few impacts and then levelled off, following closely the behaviour of the bending stiffness against the number of impacts. As for fibre damage, the vinylester impacted face specimens exhibited little damage confined to the contact area with the impactor top, while the polyester impacted face specimens showed cross-like damage patterns along fibre directions.

The impact-fatigue properties of unidirectional CF-reinforced polyetherimide (PEI) composites were evaluated [40] by subjecting standard izod impact samples to LVI at energy levels ranging from 0.16 to 1.08 J by using a pendulum impact tester. The effect of the previous LVIs on the residual impact properties of the laminates was investigated. Composite laminates were also subjected to repeated LVI tests up to fracture. Results of repeated impact study were reported in terms of peak load, absorbed energy and number of impacts. Fractographic analysis revealed fracture by primary debonding, with fibre breakage and pullout in the tensile zone, accompanied by a shear failure of fibre bundles in the compressive zone of the specimen.

The effect of solid particle erosion on the strength and fatigue properties of E-glass/epoxy composite was investigated [41]. Solid particle erosion with SiC particles of 400–500 mm in diameter was simulated on 12-ply $[45°/–45°/0°/45°/–45°/0°]_S$ composites with a constant particle velocity of 42.5 m/s and a solid particle to air volume ratio of 6 kg/m^3 at impact angles of 90°, 60°, and 30° for 30, 60, 90, and 120 s. Damaged and undamaged specimens were subjected to tensile tests while monitoring their AE activity. SEM images of the erosion-damaged specimens revealed that the same damage mechanism occurred at different impact angles. The AE stress delay parameter was used to predict the residual tensile strength of erosion-damaged composites. Tension-tension fatigue tests were performed on virgin specimens and specimens exposed to erosion damage for 60 s and 90 s at 90° particle impact angle to observe the effects of erosion damage on the fatigue life. A modified Basquin's equation was employed to predict the fatigue life of the erosion-damaged specimens.

Azouaoui et al in their work [42] reported and discussed an experimental method for characterising the damage behaviour of glass/polyester laminated composite plates at low-energy impact-fatigue. Experimental tests were performed with increasing impact energy and increasing number of impact events. The effects of impact number, energy level and cumulative impact energy on delamination area evolution, and also on crater dimensions progress (diameter and depth) were graphically presented. The delamination behaviour and crater dimension in the laminated specimens depended largely upon the level of incident energy. Well-defined impact-fatigue (E–N_f) behaviour was obtained, showing an endurance limit at above 10^4 impact cycles.

The response of adhesive joints to three fatigue regimes, namely constant amplitude sinusoidal loading (standard fatigue, SF), cyclic in-plane impacts (impact-fatigue, IF) and a combination of the two (CSIF), was investigated [43]. The samples used in the study were CFRP lap–strap joints (LSJs) bonded with a rubber modified epoxy adhesive. It was observed that fatigue fracture at very low load amplitudes occurred in IF. Two main patterns of failure were observed in SF; cohesive failure in the adhesive, which was linked to slow fatigue crack growth behaviour, and a mixed-mode failure, involving failure in both the adhesive and the CFRP. In addition, it was observed that the transition from cohesive to mixed-mode failure was accompanied by crack growth acceleration (Figure 9). In the case of IF, all failures were due to a mixed-mechanism. In the combined standard and impact-fatigue it was seen that the introduction of a relatively small number of impacts between SF blocks drastically changed the dynamics of fatigue crack propagation, increasing the crack growth rate. Furthermore, the cavitation of rubber particles in the adhesive was affected by the repetitive impact loading, which was seen as evidence of active toughening.

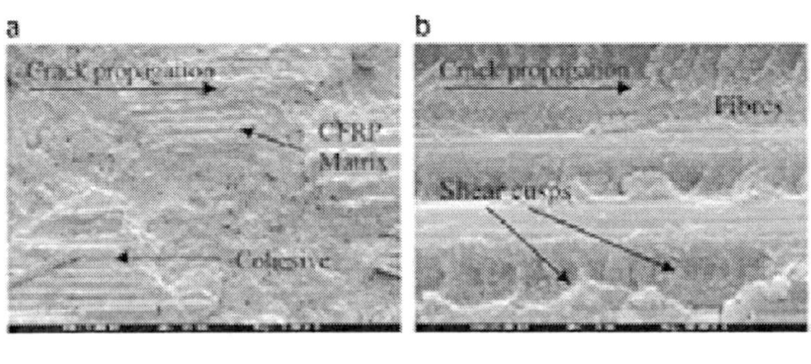

Figure 9. Impact-fatigue crack growth: failure modes [43].

The effects of thermal cycles on the impact-fatigue properties of unidirectional carbon fibre reinforced polyetherimide (PEI) matrix composites were investigated [44]. During the thermal cycles, the samples were immersed into boiling water (100 °C) and subsequently to ice water (0 °C), 50, 200 and 500 times. The changes in the viscoelastic properties of the composites were investigated by means of dynamic mechanical thermal analysis (DMTA). At the second step, thermally cycled composites were subjected to repeated impact loadings, with different impact energies. Instrumented impact test results were presented as a function of force, energy and deformation. SEM studies were performed in order to understand the morphology of fractured samples after impact-fatigue loading. As was expected, the number of thermal cycles and applied impact energy of the impactor were found to significantly influence the fracture morphology of repeatedly impacted laminates.

A damage variable (damage degree DD) representing the ratio between the absorbed energy and the impact energy was introduced in 1998 by Belingardi and Vadori to assess damage accumulation caused by LVI. More recently, repeated impact tests carried out on thick laminates pointed out the significance and extent of the penetration process in thick laminates, which DD was unable to monitor; by definition, it reached the value of one at penetration to remain unchanged over the entire penetration process. In a recent work [45], a new damage variable (damage index DI) was proposed to overcome the shortcomings of DD with respect to thick laminates. By introducing the displacement of the impactor into the definition of the DI, the depth of the penetration process was taken into account. Normalization by the displacement of the quasi-static perforation test allowed for a non-dimensional damage variable which reached the value of one at complete perforation. The validity of the approach was demonstrated against impact data obtained for different fibre-matrix architectures and laminate thickness. The DI could effectively differentiate between penetration and perforation thresholds, increasing monotonically within the range of the penetration process. In particular, the DI increased linearly with the impact energy up to penetration and rose abruptly within the range of the penetration process. Results for repeated impact tests proved that initially DI increased almost linearly impact after impact, due to a steady accumulation of damage. A few impacts before perforation a sudden rise in the DI value pointed out a change in the rate of damage accumulation.

The main aim of Casas-Rodriguez et al [46] was to investigate the behaviour of adhesively bonded CFRP joints subjected to cyclic LVI and compare it with specimens tested in standard constant amplitude sinusoidal fatigue. It was seen that the accumulated energy associated with damage in impact-fatigue was significantly lower than that associated with similar damage in standard fatigue. It was also found that the mechanisms of failure were very different for the two loading regimes. For both types of loading, fracture initiated in the adhesive layer and then propagated into the 0° ply of the composite adjacent to the adhesive layer. However, the fracture surfaces after impact-fatigue were generally less uniform and exhibited more signs of high rate or brittle fracture than those observed in fracture surfaces after standard fatigue testing. A parametric approach was proposed to characterise damage in standard and impact-fatigue and it was shown that crack velocity, accumulated absorbed energy and normalised maximum force were all useful parameters for characterising damage evolution. Crack propagation rates tended to be higher in impact-fatigue and damage occurred at significantly lower energy levels than that required for similar damage in standard fatigue. This led to the conclusion that impact-fatigue is a potentially dangerous form of loading, both for adhesives and polymer composite materials [47].

Impact-fatigue properties of unidirectional CF reinforced polyetherimide (PEI) composites were investigated [48]. Repeated LVI was performed by using a pendulum type instrumented impact tester at energy levels ranging from 0.54 J to 0.94 J. The results of the repeated impact study were reported in terms of peak load (F_{max}), absorbed energy (E_{max}) and number of repeated impacts. The life time of the composite materials was correlated to repeated impact loadings using an analytical model.

The response of natural fibre composites on the impact-fatigue loading has also been studied. An impact-fatigue study was made for the first time [49, 50, 51] on 35% jute/vinylester composites containing both untreated and alkali treated fibres. Longer alkali treatment removed the hemicellulose, improved the crystallinity and gave better fibre dispersion. The flexural strength of the composite made from treated fibres was superior. 4 h alkali treated jute fibres gave the optimum combination of improved interfacial bonding and fibre strength properties. However this was not reflected upon their impact-fatigue behaviour. On the contrary, the composites reinforced with 8h alkali treated fibres displayed superior impact-fatigue properties. SEM micrographs revealed that the fibres suffered catastrophic fracture with microfibrillar pull-out at some places, which improved the fatigue performance of the composites.

Hemp fibre reinforced unsaturated polyester composites (HFRUPE) were subjected to LVI tests in order to study the effect of non-woven fibre reinforcement on their impact properties [52]. HFRUPE composites specimens of various fibre volume fractions (V_f) were prepared and their impact response was compared with samples containing an equivalent fibre V_f of chopped strand mat E-glass fibre reinforcement. Post-impact damage was assessed using SEM. A significant improvement in load bearing capability and impact energy absorption was found following the introduction hemp fibre as reinforcement. The results indicated a clear correlation between fibre volume fractions, stiffness of the composite laminate, impact load and total absorbed energy. Unreinforced unsaturated polyester control specimens exhibited brittle fracture behaviour with a lower peak load, lower impact energy and less time to fail than hemp reinforced unsaturated polyester composites. The impact test results showed that the total energy absorbed by 0.21 V_f (four layers) of hemp reinforced specimens was comparable to the energy absorbed by the equivalent V_f of chopped strand mat E-glass fibre reinforced unsaturated polyester composite specimens.

Chapter 3

FATIGUE MODELLING – LIFE PREDICTIONS

A methodology which combines relatively short-term fracture mechanics data obtained from experimental measurements with a FEA of the component or structure was proposed for predicting the fatigue life of fibre-composite components and structures [53]. The approach was used to study the damage growth due to cyclic fatigue of I-beams. The beams were manufactured using CFRP and contained a 60 mm diameter notch in the web. Experimental work showed that the development of significant damage was limited to the region of the material in the web around the 60-mm diameter notch. A significant amount of matrix micro-cracking damage occurred within the first 0.5×10^6 fatigue cycles, mainly in the +45° plies and 0° plies, in which the fibres were oriented at 90° and at 45°, respectively, to the local tensile stress. This matrix cracking eventually led to limited delamination after approximately 0.5×10^6 cycles which took place mainly along the global +45°/–45° ply interfaces. Limited delamination was also observed along the +45°/0° ply interface. The growth of these two types of damage eventually led to fibre fracture which was the final cause of structural failure of the web material, and hence of the I-beam. The model concentrated upon modelling these types of damage mechanisms. The agreement between the results from the theoretical model and the experiments was good, especially when taking into account that there were no "adjustable factors" involved in the modelling.

The effect of matrix toughness on the fatigue behaviour of cross-ply reinforced composites was discussed by Gassan and Dietz [54]. Brittle and impact-modified epoxy resins containing commercial E-GF with a good fibre matrix adhesion were used. Load-controlled tension – tension fatigue tests with different applied maximum loads between 150 and 330 MPa were performed. The damage, as measured by stiffness reduction, was more significant for composites with a brittle matrix. The energy loss was shown to be sensitive to the matrix toughness and was higher for impact-modified resins under a given strain amplitude. A non-linearity in the S/N curve was observed for the impact-modified composites, while the S/N curve for the brittle ones was linear when plotted on a semi log-scale. Furthermore, the brittle composites showed an extensive region of final damage propagation, whereas the impact modified composites failed in a sudden death manner.

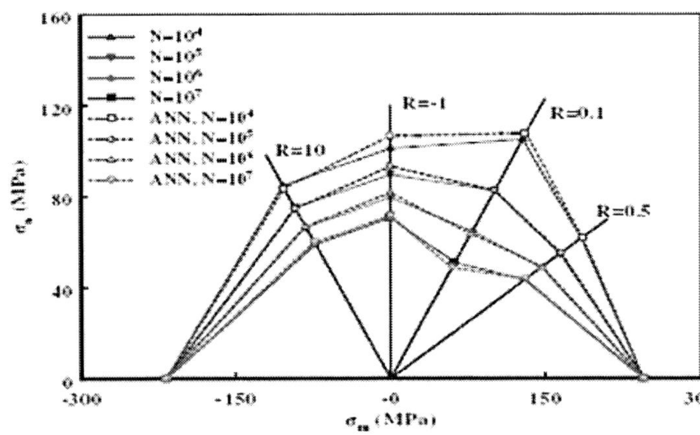

Figure 10. Constant Life Diagram (CLD) for on-axis multidirectional laminates: Artificial neural network predictions versus experimental data in the range 10^4-10^7 loading cycles [55].

An artificial neural network (ANN) was shown to be an efficient tool for modelling fatigue life of multidirectional composite laminates made of GFRP composite materials and tested under constant amplitude loading patterns [55]. The modelling efficiency of the network was satisfactory for both on- and off-axis coupons' life, irrespectively of the test conditions. Tension–tension, compression–compression and even tension–compression loading patterns were investigated and the modelling accuracy of the proposed ANN model was validated. The main benefit of this modelling tool was that only a portion, in the range of 40–50%, of the experimental data was needed for the whole analysis. Thus, expensive and time consuming tests required for the compilation of S-N curves could be significantly reduced without significant loss of accuracy. The neural network method was applied using experimental data from two different material systems and proved that constant life diagrams (CLDs) are very useful for the design of structures loaded under variable amplitude loading spectra. CLDs can be efficiently constructed in conjunction with the proposed ANN model using a smaller set of experimental data than that needed for the construction of CLDs conventionally (Figure 10).

Delamination Propagation under Fatigue – Models of Prediction

A critical review of the published experimental research was presented in [56] concerning delamination onset and growth in composite laminate interfaces of different lay-ups under single-mode loadings. It was found that, depending on the loading mode and interface lay-up, the traditional fracture toughness characterization by unidirectionally reinforced composite tests can lead to marked under- or overestimation of material resistance to crack growth. Empirical models of fracture toughness as a function of delamination front orientation with respect to reinforcement directions of the adjacent laminae were validated and their applicability range was established. The delamination onset and growth in composite laminates was also characterized in two phenomenological works by Ramkumar [57, 58]. Fracture mechanics was used to model delamination growth during static loading [2, 59, 60, 61, 62]. Models for describing delamination growth under cyclic loading were proposed [63, 64]. Trethewey et al [65] studied the delamination propagation under pure Mode II conditions.

Behesty et al [66] elaborated on earlier work about the development of a life-prediction method for CF/epoxy laminates. In order to compare the behaviour of a number of different CFRP laminates already studied, further constant-life fatigue data were obtained for a CFRP composite and a GFRP laminate of similar construction – a 16-ply $[(\pm45,0_2)_2]_S$ lay-up. Fatigue tests were carried out in both virgin condition and after LVI damage. The analysis of the new data and the re-assessment of the older data base, led to the appropriate modification of the constant-life model. A prediction procedure for the fatigue response of composite materials in the virgin and impact-damaged conditions was presented, based only on the tensile and compressive strengths of the composite in question. The model was equally applicable to both GFRP and CFRP, despite the fact that the fatigue response of the GFRP laminate was different from that of the equivalent CFRP material.

Shen et al [67] dealt with the computational modelling of delamination and the prediction of delamination growth in laminated composites. In their analysis of post-buckled delaminations, an important parameter was the distribution of the local strain-energy release rate (SERR) along the delamination front. A study using virtual crack closure technique was performed for three-dimensional FEA models of circular delaminations embedded in woven and non-woven composite laminates. The delamination was embedded at different depths through the thickness of the laminates. The issue of symmetry in boundary conditions was discussed. It was found that fibre orientation of the plies in the delaminated part played an important role in the distribution of the local SERR. This implied that the popular use of quarter models in order to save computational effort is unjustified and may lead to erroneous results. A comparison was made with experimental results and growth of the delamination front with fatigue cycling was predicted. A methodology for the prediction of delamination areas and directions using evolution criteria derived from mechanical tests was also described. It was found that evolution criteria based on components of the SERR predicted the rate of delamination growth much better than evolution criteria based on the total SERR.

A thermodynamically consistent damage model was proposed for the simulation of progressive delamination in composite materials under variable-mode ratio [68]. The model was formulated in the context of Damage Mechanics. A novel constitutive equation was developed to model the initiation and propagation of delamination. A delamination initiation criterion that evolved from the Benzeggagh – Kenane propagation criterion was proposed to assure that the formulation can account for changes in the loading mode in a thermodynamically consistent way. The formulation accounted for crack closure effects to avoid interfacial penetration of two adjacent layers after complete decohesion. The model was implemented in a finite element formulation, and the numerical predictions were compared with experimental results obtained in both composite test specimens and structural components (Figure 11).

Kim and Hwang [69] proposed an impact response model that can be used to compute the fatigue damage. The prediction of fatigue damage can be made through investigation of impact response. The stiffness reduction induced in composite laminates by matrix cracking or delamination is an important measure of fatigue damage. Additionally, stiffness reduction affected impact response. The authors identified a correlation between fatigue and impact damage manifested as stiffness degradation. Impact induced stiffness reduction was investigated using simplified methods and the predicted impact force histories were compared with numerically calculated ones using FEA.

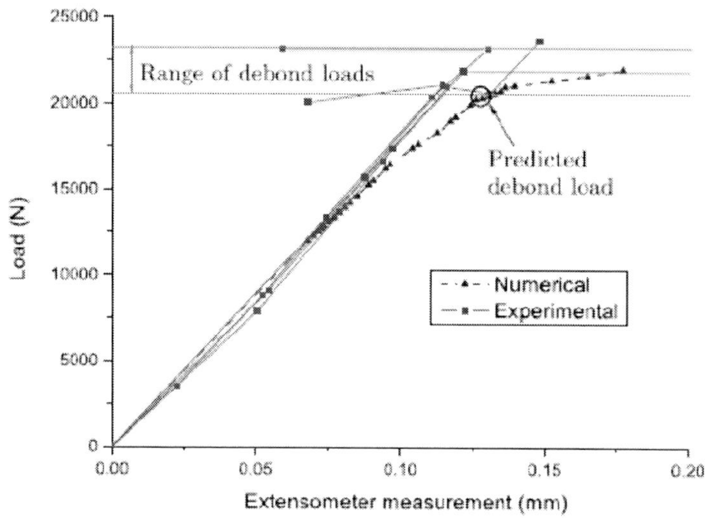

Figure 11. Experimental and numerical load-displacement relations for fatigue after impact [68].

Chapter 4

NDT INSPECTION OF INITIAL DAMAGE AND DAMAGE PROPAGATION

ACOUSTOGRAPHY

Acoustography provides a very powerful means of monitoring damage in composites in real time during both static loading and long-term dynamic fatigue tests. The integration of an acoustography ultrasonic imaging system with a mechanical testing machine was described [70, 71, 72]. The apparatus was designed for real-time ultrasonic imaging of impact damage growth in composite material samples subjected to either static or fatigue loading (Figure 12). Experimental results of the damage growth up to failure of an aerospace standard CF composite subjected to compressive cyclic loading were presented. Experimental results for static loading revealed a reversible increase and decrease in the damage area, indicating an opening and closing of delamination defects, which pointed to the possibility of an under-estimation of defect size by the conventional ultrasonic testing of unloaded components. Results presented [71] showed the damage-area growth during fatigue cycling under high compressive loads. After an initial small enlargement, damage grew at a constant rate until the third stage was reached when there was further growth at an increasing rate to final failure. A "fatigue limit" was also observed.

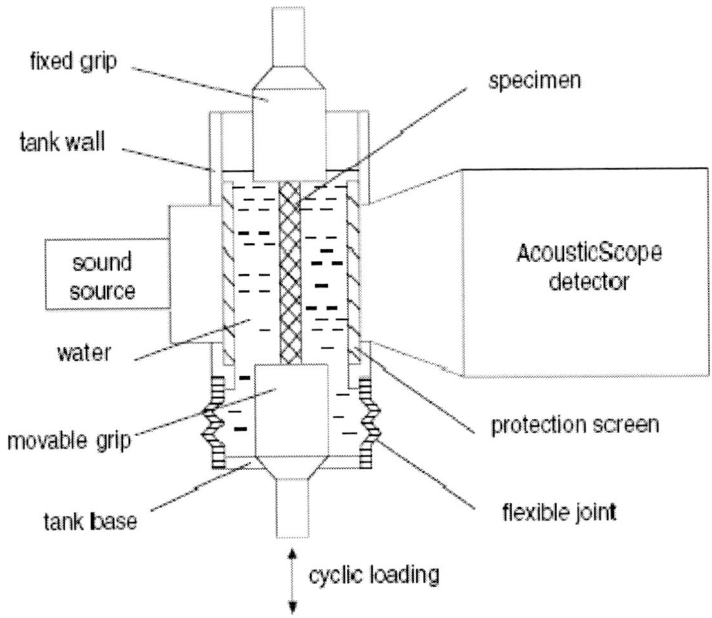

Figure 12. Acoustography: Schematic diagram of the system [70].

Acoustic and Lamb Waves' Methods

The work of Grondel et al [73] was devoted to the development of a health monitoring system designed for aerospace applications. These applications were concerned with the detection of damaging impacts and debonding between stiffeners and composite skins, since these are the major causes of in-service damage of aircraft structures. The chosen health monitoring system was first based on the excitation and reception of Lamb waves along the structure by using thin piezoelectric transducers (active mode) and secondly on a continuous monitoring taking the same transducers used as acoustic emission sensors (passive mode). The composite specimen used was consistent with an aircraft wing-box in terms of structure and loading. Several impacts with incrementing energy were applied on the composite specimen. In passive mode, the study showed the ability of using the acoustic signature of an impact to detect possible damage. Moreover, the damage initiation in the case of damaging impact was confirmed in active mode. Further measurements during fatigue testing were performed. The ability of the system to monitor disbond between the stiffener and the composite skin was demonstrated and the sensitivity of the health monitoring system to the disbond growth was further evaluated.

Toyama and Takatsubo [74] proposed an inspection technique using Lamb waves to detect impact-induced delamination in composite laminates. The technique consisted of two line scans. The first scan measured the arrival times of the transmitted S_0 mode along the 0° direction to detect delamination and evaluate its size. The second scan measured the maximum amplitude of the earliest wave packet in a line, including the longest delamination, to locate its edge. The technique was performed on impacted CFRP cross-ply laminates. A remarkable decrease in the arrival times due to the delamination was detected, and the delamination length was calculated using a simple model for Lamb-wave propagation. Furthermore, the delamination edge was located as a sudden decrease in the amplitude. The technique enabled the detection of the delamination and the evaluation of its size and location using only two scans.

The article of Monnier [75] used an in situ NDI method for the impact damage monitoring of carbon/epoxy laminated composites. This structural health monitoring (SHM) method aimed at improving the operational safety of the structure using integrated piezoelectric sensors. The propagation of Lamb waves inside a stiffened aeronautical structure, and the definition of associated data processing systems were described. The detection of realistic defects was made feasible by the definition of a damaging parameter, based on the real time analysis of the ultrasonic signatures transmitted inside the structure.

An LVI simulation by loading at the centre of quasi-isotropic CF reinforced epoxy resin laminate was presented in order to characterize damage development [76]. Acoustic emission (AE) was used to detect damage accumulation. It was found that the drops in the load – displacement curve were due to delamination and fibre failure. The damage induced by the 7 mm radius indenter was much higher than the one induced by the 4 mm radius as a higher fraction of fibres failed, further reducing the residual strength. The tensile strength after indentation was also investigated and correlated with AE parameters. In the tensile test the residual strength of the $[0/90/+45/-45]_s$ lay-up was higher as the delamination and splitting of the external lamina prevented fibre failure during transverse loading. A FEA model for the prediction of both the first ply failure (FPF) and the ultimate ply failure (UPF) of the laminate was built using the ANSYS® software and correlated well with experimental findings.

Electrical Measurements

Self-sensing of damage by measurement of the DC electrical resistance or potential away from the damaged region was demonstrated [77] in quasi-isotropic continuous CF epoxy-matrix composite laminates under impact at energy up to 5 J. The through-thickness potential drop was substantial up to 240– 480 mm (at 0.25–99 mA correspondingly) in the longitudinal direction from the position of through-thickness current application, due to current spreading in the longitudinal direction. A model for the current spreading was also provided. The fractional change in resistance resulting from damage decreased with increasing distance from the point of impact (diameter of indentation up to 3.5 mm and depth of indentation up to 0.16 mm), but was detectable even at a distance of 150 mm from the point of impact. Both the through-thickness resistance and the oblique resistance were effective indicators. The ability for the resistance measured away from the damaged region to indicate damage in the damaged region was attributed to the considerably lower electrical resistivity in the longitudinal than in the through-thickness or oblique direction in the composite.

Drop impact damage of continuous carbon fibre epoxy–matrix composite laminates, was studied [78] by electrical resistance measurement, and was proved to be more sensitive than the ultrasonic method. The oblique resistance at an angle between the longitudinal and through-thickness directions was more effective than the surface longitudinal resistance in indicating damage, particularly in the interior of the laminate (Figure 13). The oblique resistance values from longitudinal segments of a specimen could not be integrated, but the surface resistance values were. In the case of a unidirectional composite, electrical contacts at 45° from the longitudinal direction in the plane of the laminate were more effective than those at 90°. Even minor damage associated with negligible indentation was sensed. The spatial distribution of damage was also studied.

The method of sensing impact damage in carbon fibre polymer-matrix structural composite by DC electrical resistance measurement was evaluated by measuring the resistance of the impacted surface [79]. The resistance was obtained by using the four-probe method as a more sensitive, more precise (less data scatter) and more accurate indicator of composite damage than that obtained by using the two-probe method. The data scatter was low for both four-probe and two-probe resistance for impact energy up to 5 J, but it was lower for the four-probe resistance than the two-probe resistance and increased with damage. The four-probe resistance, of the 8-lamina composite, increased upon impact, and the fractional increase diminished as the distance from the point of impact increased. The four-probe resistance of the 24-lamina composite increased upon impact for the specimen segment containing the point of impact, but decreased slightly upon impact for the segments within 20 mm from the point of impact. The two-probe resistance decreased less upon impact than the four-probe resistance.

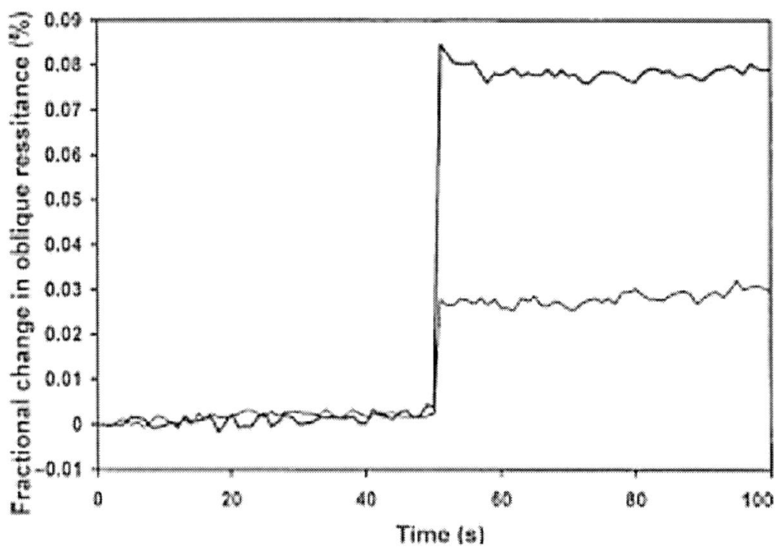

Figure 13. Fractional change in oblique resistance upon impact at 0.73J (thin curve and 3.63 J (thick curve) [78].

Self-sensing damage, as obtained by electrical resistance measurements, was found to be effective in a carbon fibre polymer–matrix composite cylinder made by filament winding [80]. Resistance was measured in the axial, radial, oblique, and circumferential directions by using circumferential or axial electrical contacts on the outer and/or inner surfaces of the cylinder. Minor damage upon drop impact at 10 J or below caused the radial resistance to decrease irreversibly, whereas major damage upon drop impact above ≈10 J caused the radial, oblique, and axial resistances to increase irreversibly. The circumferential resistance ratio i.e. the ratio of the circumferential resistance of a damaged area to that away from the damaged area was most sensitive and it increased monotonically with impact energy over 1.4 J.

The electrical resistance method and the electrical potential method for composite damage self-sensing were compared for the case where the direction of the applied current and the direction of the measured potential gradient were parallel [81, 82]. The electrical potential method involved measuring the potential at a distance from the line of current application and was effective for damage sensing of carbon fibre polymer–matrix composites when the distance in the through-thickness direction was sufficiently small or else for thin laminates up to eight-plies. However it was ineffective for 16 ply laminates i.e. for a 2.1 mm thickness. On the other hand, the electrical resistance method, which involved measuring the potential on the line of current application, was not limited by the thickness of the laminate. However, it suffered from current path distortion upon damage and the consequent reduced sensitivity for damage sensing. Interestingly enough, the measured resistance was reported to decrease with increasing damage for the region of the composite not containing the point of impact. However, this effect was counterbalanced by the resistance increase for the regions containing the point of impact.

The self-sensing of flexural strain and damage was demonstrated [83] in carbon fibre polymer-matrix composite by measuring the DC electrical resistance. Upon strain in the elastic regime, the compression surface resistance decreased reversibly (due to increase in the current penetration), while the tension surface resistance increased reversibly (due to decrease in the current penetration), and the oblique resistance increased reversibly. Upon minor damage the oblique resistance after unloading decreased, the oblique resistance decreased with load application near the onset of loading and the curve of the oblique resistance or the resistance of the tension or compression surface vs. deflection became nonlinear. Upon major damage, all resistances abruptly and irreversibly increased. The onset of damage as manifested by the increase of surface and oblique resistance occurred earlier for the compression surface than for the tension surface. The surface resistance was a superior indicator for strain, whereas the oblique resistance was a superior indicator for damage.

Shen et al [84] used beam-type specimens with and without delamination damage to carry out numerical analyses for the application of the four-probe electrical resistance measurement. The validity range and the applicability of the method were studied. It was found that the four-probe resistance method was valid only when the through-thickness conductivity was comparable to or larger than the longitudinal conductivity. For the potential method, which measured directly the voltage values between the voltage contacts, the results showed that the percentage change in damage-induced voltage drop between a pair of voltage contacts was not consistent with the percentage change in resistance. The underlying reason was that the damage-induced voltage change depended on the location of the applied current, while the resistance change did not.

Todoroki et al [85] employed an electrical resistance change method for monitoring delamination. They found both experimentally and analytically that the electrical resistance change method using response surfaces was very effective in identifying delamination in CFRP laminates. The effect of the electrode spacing was investigated using Finite Element Analysis (FEA). Five types of spacing were analyzed for two fibre volume fractions. Cross-ply beam type specimens were adopted for the analyses. It was shown that the effect of the spacing depended on the fibre volume fraction. For laminates of high fibre volume fraction, dense spacing was required to obtain an accurate estimation of delamination location and length.

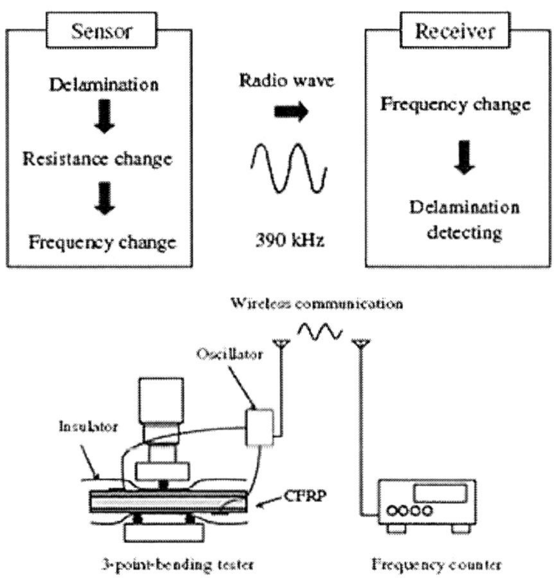

Figure 14. Wireless delamination detection system using electrical resistance and oscillating frequency change [85].

Matsuzaki and Todoroki [86] proposed a wireless system for detecting delamination of carbon/epoxy composites to overcome the difficulties of in-service delamination detection for rotating composite components like helicopter and wind turbine blades. In this system, a tiny oscillation circuit was attached to the composite component. When delamination of the component occurred, the electrical resistance changed causing a change in the oscillating frequency of the circuit (Figure 14). Since this system used the composite structure itself as a sensor and the oscillating circuit was very small, it was applicable to rotating components. The electrical resistance change and oscillating frequency change due to delamination were experimentally measured for carbon/epoxy specimens. The effect of temperature variation was also measured. The wireless method was found to successfully detect delamination, and to estimate its size. The effect of temperature change was minimized by means of a temperature compensation circuit.

Ueda and Todoroki [87] employed a two-stage electric potential change method (EPCM) to identify delamination. Delamination is usually estimated using response surfaces, which requires numerous experiments. An equivalent electrical conductivity method was introduced to reduce this number. Delaminations were successfully estimated using response surfaces based on FEA results for equivalent electrical conductivity.

The electrical potential technique was applied [88] to detect and locate impact damage in CFRP plates. The potential field across the surface of the laminates was measured using arrays of electrical contacts. A constant current of 100 mA was applied to the plate, and the changes in the potential distribution resulting from impact damage were measured. These results were compared to those calculated using 3D FEA simulation of current flow in CFRP laminates. The results showed good agreement between simulation and experimental results and could allow for detailed analytical studies of optimum network and current input configurations for the practical realization of a damage sensing system. A comparison of the potential distribution on the top surface for damaged and undamaged laminates showed a substantial difference in the potential field around the impact damaged area.

Quasi-isotropic samples of CFRP composite 2 mm thickness were manufactured and instrumented with an array of electrodes [89]. The electrodes were used to introduce a DC current and to measure the resultant potential distribution on the top and bottom surfaces of the laminate. Instrumented samples were impacted at energies between 2 and 12 J, and the potential fields were measured before and after impact. The impact damage caused significant changes in potential values. Indications of the extent and severity of the impact damage could be best obtained by representing the data as contours of equi-potential change. There was excellent correlation between regions of potential change and impact damage as measured by ultrasonic C-scan. The results of FEA calculations of potential change distribution on the surface of current carrying quasi-isotropic laminates containing delaminations of a range of sizes and locations were qualitatively similar to those obtained experimentally, but differed in detail. The differences between real and simulated impact damage were discussed to assess the performance of a damage detection system based on equi-potential contour mapping.

Embedding Optical Fibres

Kuang and Cantwell [90] presented the results of a study involving the use of conventional glass optical fibres for detecting impact damage in a thermoplastic-based Fibre-Metal Laminate (FML) system. Although a number of researchers have investigated the use of intensity-based optical fibre sensors for detecting impact damage in composites, almost all of the published literature has focused on detecting damage in plain thermosetting polymer composites. This study investigated the possibility of using intensity-based optical fibres for detecting impact damage in thermoplastic-based FML systems. The results suggested that for higher fracture detection sensitivity, optical fibres should be embedded close to the impact surface in order to be sufficiently sensitive for energies of 1 to 2 Joules in these hybrid structures. Although the optimum choice of optical fibre and its location should be re-established for other types of material system, this study highlighted some important observations for the future development of damage detection systems for use in tough composite materials and fibre metal laminates.

Step-index multi-mode optical fibres were embedded in a thermoplastic glass fibre polypropylene (GF/PP) [91]. Three types of commercially available optical fibres were investigated to evaluate their potential for detecting impact-induced damage in thermoplastic-based composite structures. Preliminary findings confirmed the feasibility of using these inexpensive multi-mode optical fibres as damage sensors in high-performance composite materials. When embedded between the uppermost ply of a cross-ply unidirectional $[0/90]_s$ laminate, impact energies as low as 0.8 J could be detected. The optical fibres were embedded at different locations within the composite structure to investigate the effect of embedment locations on their sensitivity for detecting impact damage.

Kister et al [92] reported the use of conventional reinforcing E-Glass Fibres that were employed as light guides. These reinforcing fibre light guides were used to detect damage induced in the composite by impact, indentation and flexure. The glass fibres were converted into light guides by applying an appropriate cladding material. The coating resins used in this study consisted of an epoxy- and a polyurethane-based resin system. These self-sensing fibres or reinforcing fibre light guides were either surface mounted or embedded at two specified locations within 16-ply glass fibre-reinforced epoxy prepreg composites. The data generated in this study demonstrated that the self-sensing concept could be used to study *in situ* and in real time the failure processes in Glass Fibre reinforced composites. A detailed study was also undertaken to characterize the various failure modes observed when the composites with the self-sensing light guides were subjected to impact, indentation and flexural loading. The damaged areas in the composite were easily located by means of the "bleeding" light emanating from the broken self-sensing E-GFs (Figure 15).

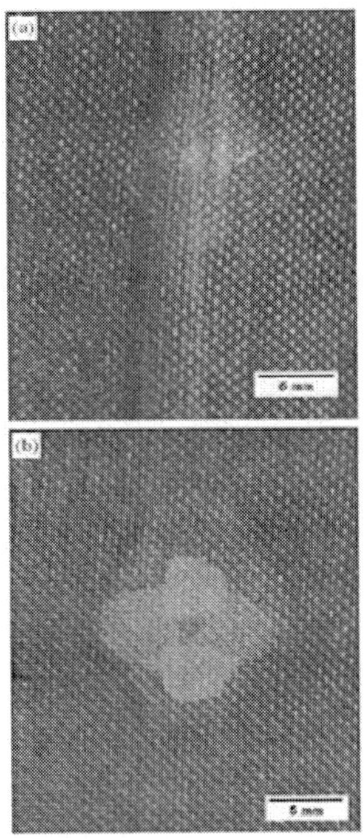

Figure 15. Delamination of a composite subjected to 2 J impact, were the glass fibre reinforcement is used as a light guide (a) front face, (b) back face [92].

Small-diameter fibre Bragg grating (FBG) sensors were employed [93] for the monitoring of delamination induced by LVI. The FBG sensors were embedded into CFRP laminates (Figure 16). Using a drop-weight impact tester, impact loading was applied to the laminates at four impact energy levels. After the impact tests, the internal damage including delaminations was observed by ultrasonic C-scan, and the reflection spectra from the embedded FBG sensors were measured. The form of the spectrum changed significantly depending on the delamination size. Furthermore, the spectra were calculated theoretically for confirmation of the measured spectra. Since the change in the measured spectrum was consistent with that for the calculated spectrum, the relationship between the delamination size and the form of the spectrum could be clarified. From the results, the present method using small-diameter FBG sensors was found to be effective for the monitoring of the delamination.

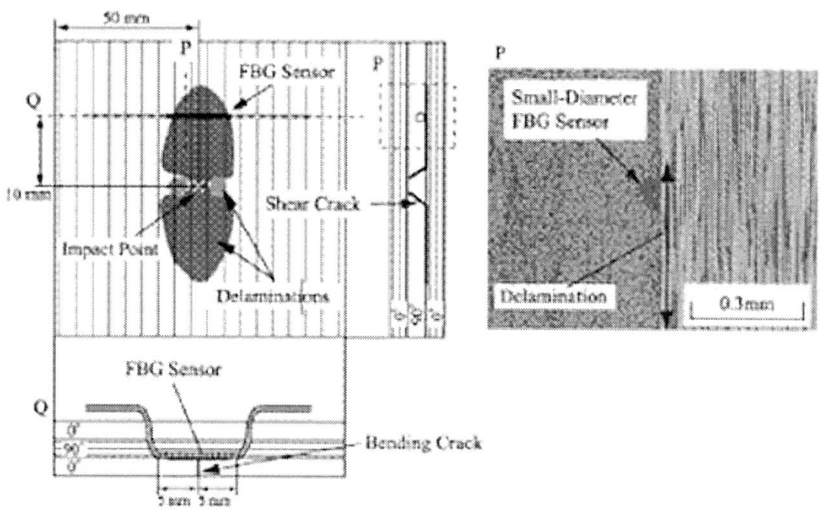

Figure 16. Embedding a small diameter FBG sensor into the test specimen to detect delamination [93].

The purpose of Silva et al [94] was to quantitatively evaluate the effect of embedding optical fibres on the mechanical behaviour of a CF/epoxy composite in order to assess whether their presence could potentially degrade the mechanical performance of the host material. The existing literature on this subject is not conclusive about the nature and intensity of this effect. Three kinds of mechanical tests were performed in this work: impact tests, static flexural tests, and fatigue tests. The results pointed to a possible detrimental effect related to the presence of optical fibres which was different in nature and intensity for each of these tests. The mechanical behaviour in static loading conditions was not significantly affected by the presence of the optical fibres. However, impact and fatigue performance was strongly affected albeit in different ways. Based on these results, the possible failure mechanisms that can explain the detected differences were discussed.

Damage in unidirectional CF composite resulting from both LVI and HVI was evaluated using embedded FBG sensors, C-scan and microscopic analysis [95]. It was found that the FBG sensors located 10 mm from the impact site could detect residual strains from a 0.33 J (1.3 m s^{-1}) impact which was not detectable by C-scan or visual inspection. The measured residual strain increased with impact energy and damage changed from matrix cracking to severe delaminations. HVI (225 m s^{-1}, 11 J) resulted in test panel perforation and delaminations. FBG sensors located within a distance of 2–3 the damage radius detected residual strain from the impact. An array of embedded sensors could possibly be used to identify the site of both LVI and HVI and predict the damage from the response of the adjacent sensors providing the sensors are located sufficiently close to the impact site.

FBG sensors were applied to durability tests of a composite wing structure in order to verify its health monitoring capabilities for long-term use [96]. The durability tests included drop-weight impact tests and two periodic fatigue tests based on the design service life of the aircraft. The seven FBG sensors were installed to the surface of the test panel for the monitoring of impact and fatigue damage. The impact damage was detected by using the spectrum change of the sensor output. The strain changes during the fatigue tests were measured by the wavelength shift of the sensor output. The damage evolutions in the test panel were also evaluated by using various NDI technologies, i.e. AE, ultrasonic C-scan and pulse thermography. Compared with results of the other non destructive techniques, the FBG sensors could provide valuable information for monitoring the structural integrity of the composite wing structure. As a result, it was confirmed that FBG sensors were capable for the long-term health monitoring of large-scale composite structures.

A study of the response to LVI and perforation of two-dimensional and three-dimensional woven composites was undertaken utilizing local measurements of post impact residual strain measured by surface mounted and embedded optical fibre Bragg grating (FBG) strain sensors [97]. Concurrent global measurements of contact force, dissipated energy, and point of impact displacements were undertaken. The study delineated five distinct regimes ranging from initial impact to complete penetration. Sensor and host damage were separated by signal intensity and the evolution of Bragg peaks due to repeated impact loads. The results indicated that a local-global framework could be used to monitor damage progression in different host materials, and hence it could be potentially used to mitigate damage effects. The impact strain relaxation occurred in two forms: a global relaxation of the average strain over the grating seen by uniform Bragg peak shifts, and a reduction in the strain fields' non-uniformity over the grating length as indicated by the non-uniform reductions in the peak bandwidth around the Bragg wavelength. The measured residual impact strain relaxation reached a maximum of one-third of the peak residual impact strains. Due to strain relaxation, the amount of time between impact events provided enough time for sample equilibration, which had an effect on the composites' response to additional impacts.

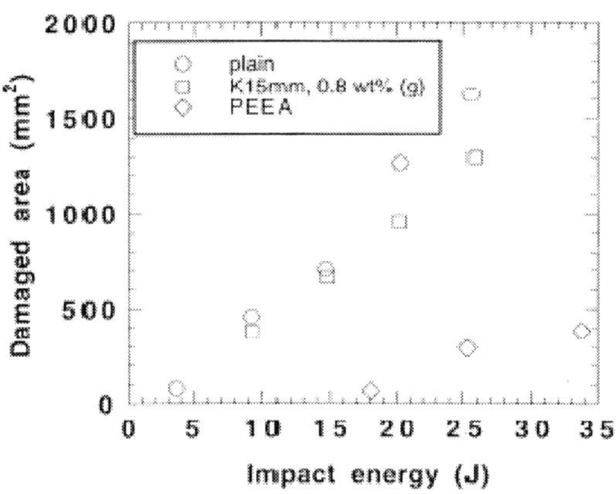

Figure 17. Total damage versus impact energy estimated by Scanning Acoustic Microscopy [98].

Imaging Techniques

LVI tests were performed to investigate the impact behaviour of CF/epoxy composite laminates reinforced by short fibres and other interleaving materials. Characterisation techniques, such as cross-sectional fractography and scanning acoustic microscopy (SAM), were employed [98] to assess quantitatively the internal damage of composite laminates at the sub-surface under impact (Figure 17). SEM was used to observe impact fractures and damage modes at the fracture surfaces of the composites. The laminates experienced various types of fracture; delamination, intra-ply cracking, matrix cracking, fibre breakage and damage depending on the interlayer materials. The trade-off between impact resistance and residual strength was minimised for composites reinforced by Zylon fibres. For composites interleaved by poly(ethylene-co-acrylic acid – PEEA) film, the deterioration of residual strength was substantial, although the damaged area was significantly reduced. Damage induced on the front and back surfaces of impact was also observed and compared for different laminates.

The rapid estimation of the damage of a composite plate was possible with thermal methods. The use of high performance cameras based on focal plane array of quantum detectors and fast post processing led to the substantial improvement of thermal methods. Post processing consisted in associating the experimental advantage of the flash method and some characteristics of periodic methods with the processing of a big quantity of data. The main idea was to use as a reference signal the spatially averaged image evolution. Some practical aspects of the method applied to the estimation of delamination characteristics were presented by Guillaumat et al [99]. The experiments were obtained from damaged composite plates (transparent fibre glass epoxy sample covered with a black painting).

Mian et al described the application of a short ultrasonic pulse in combination with infrared surface imaging to assess fatigue damage in an unnotched [100] and notched [101] graphite/epoxy composites. The technique used a short pulse of sound wave that infused into the material and resulted in frictional heating at the crack surfaces. The heating at the crack defused into the surface of the sample, and the temperature rise at the crack location was sufficient to be captured by an infrared (IR) camera. The ability of the sonic infrared imaging technique in detecting fatigue damage in composite materials was demonstrated. FEA modelling was performed to better understand the effect of the applied sound pulse at the damage locations and to interpret the vibrometer data. The same samples were also tested using a thermal wave imaging technique. In the case of notched composites the technique revealed the presence of semi-elliptical delaminations near the circular notch as modelled in the FEA. A separate FEA model was developed for transient heat transfer analysis by utilizing the calculated frictional energy as a heat source. Variation in surface temperature within the delamination regions was plotted with time, and was compared with the temperature–time plot obtained from the sonic IR technique.

The study by Meola et al [102] focused on the use of lock-in thermography (LT) for NDI evaluation of aerospace materials and structures. The experimental analysis was performed by testing several specimens, which were made of different materials employed in the fabrication of aircraft (composites, hybrid composites, sandwiches, metals) and which included the most commonly encountered types of damage (delamination, impact damage, fatigue failure). The data presented showed that LT provided useful information, which could be exploited for detection of defects and evaluation of their size, position and nature (Figure 18).

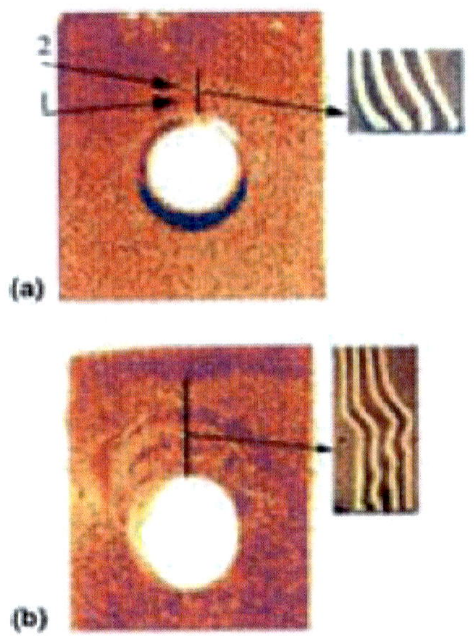

Figure 18. Lock-in thermography: Phase images and micrographs of Glare with (a) and without (b) lateral restraints [102].

A digital imaging technique for characterizing dynamic delamination in laminated composites was presented by Wosu and Hoy [103]. Perforation induced delamination surfaces were generated using a conical and protruding hemispherical penetrator applied in the thickness direction using the penetrating split Hopkinson pressure bar (P-SHPB). The studies revealed that damage layers on perforated specimens increased linearly below the perforation threshold energy and sharply after perforation.

Ultrasonics

Delamination extent is usually measured by ultrasonic C-scan, which provides an in plane view of the damaged zone. By this technique, some information on the actual damage state is lost, as it is typically made of several interlaminar cracks located at different heights along the material thickness [104]. Nevertheless, a C-scan procedure is easy to perform and relatively inexpensive and experimental evidence given that the in-plane dimensions of delamination can be correlated with the material residual strength. Furthermore, the delamination area as revealed by ultrasonic C-scan is frequently used to rank different laminates on the basis of their impact resistance.

Unidirectional graphite/epoxy composites used in the aerospace industry, were impacted by a free falling weight [105]. The subsequent changes/ degradation in elastic moduli, strength, toughness, and fatigue properties were measured after different number of impacts. It was found that for all energy levels these properties varied linearly with the number of impacts. Attenuation changes were not a good parameter for degradation estimation, since they did not incorporate the micro- and macro-cracks beyond the impact point. However, these micro- and macro-cracks had significant effect on the mechanical properties. In contrast to the attenuation, the stress wave factor, which indicated the efficiency of wave propagation along the specimen, correlated very well with degradation, and could be used effectively to measure the residual strength after impact. The ultrasonic characteristics of specimens subjected to combined fatigue and impact were also studied. Based on these experiments, it was concluded that the loss in fatigue residual life due to impact could be predicted by measuring the effects of the impact load on attenuation and the stress wave factor. It was found that the reduction in fatigue life was proportional to sudden changes in attenuation and stress wave factor. Damage accumulation models based on the Coffin-Manson equation were suggested for impact and combined fatigue and impact. It was found that residual properties and fatigue life could be estimated using these models.

The aim of Adden et al [106] was to demonstrate the possibility of fatigue damage characterization in GFRP-tube-like components by using circumferential plate waves (Figure 19). For that purpose, fatigue tests with different loading directions were conducted and the stiffness degradation was monitored. After a preset number of loading cycles, non-destructive ultrasonic tests using circumferential plate waves were performed. The correlation between damage induced amplitude changes of the plate waves and stiffness degradation in non-crimped fabric composites was discussed. The results indicated that the technique was applicable to fatigue damage assessment in complex-shaped components of composite materials and was applicable for a wide range of applications.

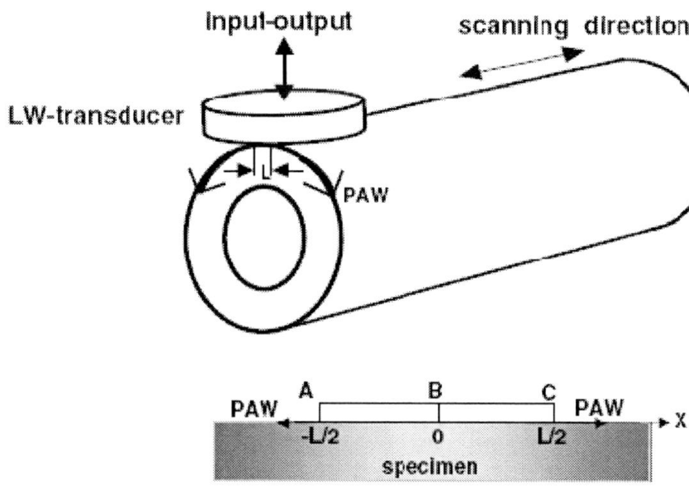

Figure 19. Schematic representation of the fatigue damage measuring device using circumferential plate acoustic waves [106].

Other Methods

Optical coherence tomography (OCT) is an emerging technique for imaging of synthetic materials. OCT is attractive because it combines high sensitivity (> 90 dB), high resolution (5 μm to 20μm), and low cost. The value of any new technology is evaluated by how well it compares with existing methods. In the work by Dunkers et al [107], impact damage of an epoxy/E-glass composite was imaged using OCT, and the results were compared with micro-focus X-ray computed tomography. This technique was regarded as a good benchmarking method for OCT because both techniques have the ability to locate features precisely and have comparable resolutions. OCT was considered to be a confocal technique so it was also compared to laser scanning confocal microscopy (LSCM). Contrast mechanisms, sensitivity, resolution, depth of penetration, and artefacts of the techniques were compared. The impact damage features revealed using OCT was also briefly discussed.

A novel experimental technique was developed for time-resolved detection and tracking of damage in the forms of delamination and matrix cracking in layered materials such as composite laminates [108]. The technique was non-contact in nature and used dual or quadruple laser interferometers for high temporal resolution (Figure 20). Simultaneous measurements of differential displacement and velocity at individual locations were obtained to analyze the initiation and progression of interfacial fracture and/or matrix cracking/delamination in a polymer matrix composite laminate system reinforced by graphite fibres. The measurements at multiple locations allowed the determination of the speeds at which the interfacial crack front (mode-I) or the matrix cracking/delamination front (mode-II) dominated. Experiments were carried out using three-point bending configurations. Impact loading was achieved using a modified Kolsky bar apparatus with a complete set of diagnostics for load, deformation, deformation rate, and input energy measurement. This technique was used to characterize the full process of damage initiation and growth. The experiments also focused on the quantification of the speed at which delamination or damage propagated under primarily mode-I and mode-II conditions. The speed of delamination (mode-I) or the speed of matrix cracking/delamination (primarily mode-II) increased linearly with impact velocity. Furthermore, speeds of matrix failure/delamination under primarily mode-II conditions were much higher than the speeds of induced delamination under mode-I conditions.

Růžek et al [109] compared the results of visual, ultrasonic C-Scan and laser shearography for impact damage assessment of sandwich panels, which were cut out from sandwich skins. Considering the reliability, simplicity and rapidity of each technique and with the digital indicator measurement as a base, the laser shearography was evaluated as the most suitable method for that purpose.

The objective of the work presented by Amaro et al [110] was to evaluate the features and capabilities of four different methods when utilized to detect and quantify impact damages on CF-reinforced epoxy composite. Experimental tests were performed on $[0/90/0/90]_{2s}$ and $[90_4/0_4]_S$ laminates, using a drop weight-testing machine with impact energy of 3 J, which corresponded to a maximum load of approximately 3000 N. The damage in the laminates was analysed by electronic speckle pattern interferometry (ESPI), Shearography, Ultrasonic testing and X-radiography. All techniques were able to detect almost all the defects, however, the interferometric methods showed some limitations. X-radiography was presented as a promising alternative technique, but was not able to localise delaminations in the through thickness direction. The ultrasonic methods, A-scan and C-scan showed to be the best solutions to inspect the samples. According to the experimental results, these techniques were able to detect and measure defect extension on whole samples with good precision.

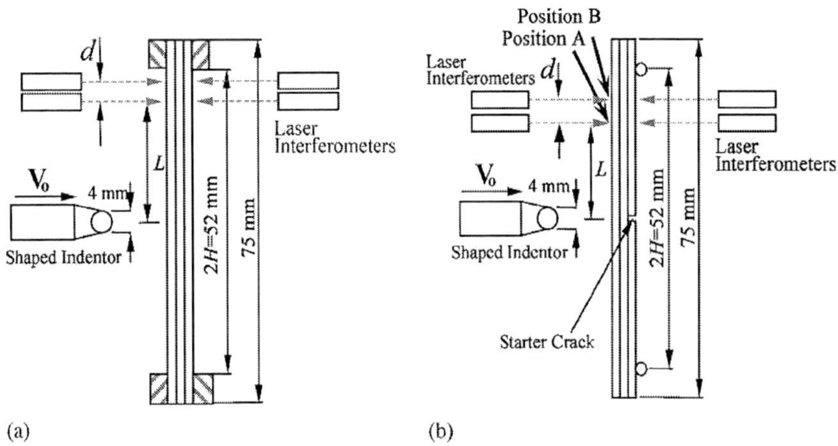

Figure 20. Non-contact detection of interfacial fracture: (a) center crack (primarily mode I) and (b) shear crack (primarily mode II) configuration [108].

The main purpose of the Corigliano et al study [111] was to assess the performance of the unscented Kalman filter for real-time identification of unknown properties of the debonding surface(s) on the basis of free-surface measurements only. The problem of impact induced delamination in layered composites was numerically modelled on the basis of the following assumptions: the layers are elastic; softening interface laws model the progressive interlaminar de-cohesion due to delamination. An explicit dynamic finite element procedure was used for the step by step analysis. Results concerning impact on a GFRP composite were discussed.

Hatsukabe et al [112] presented their latest progress in the studies on SQUID-NDE for CF composites (CFCs), including CFRPs and CF reinforced carbon matrix composites (C/Cs). A new SQUID-NDE system for the CFCs was developed, utilizing a pulse tube cryocooler for more practical and easy-handling use. Electromagnets using U-shaped ferrite cores have been introduced to the system in order to induce enough eddy current density in the relatively low-electric-conductivity CFCs. By using this system, the detection of deep-lying defects in thick and/ or multi-layer CFCs specimens was demonstrated by partly including a metal layer upon or below the CFCs layer. CFRP plates and multi-layer C/C plates with total thickness of 20 mm having deep-lying slots at various depths were prepared. The slots at 15 mm below the surface in the CFRP plate and the multi-layer C/C plate were successfully detected. Defect detection on more 'realistic' multi-layer structures, including aluminium plates upon/below C/C plates, was also attempted.

Chapter 5

POST – IMPACT BEHAVIOUR UNDER STATIC AND CYCLIC LOADING

In the following paragraphs, a detailed presentation of research works conducted to investigate the post-impact behaviour of composite laminates under static or cyclic loading will be given. The experimental studies and their conclusions are summarized.

The effect of prior damage on the tensile strength and toughness of a number of CFRPs and GFRPs angle-ply laminates was investigated [113]. The methods of introducing this prior damage were tension-tension fatigue, transverse compression, repeated impact, and stress corrosion. The effects of damage were assessed by measuring the un-notched and notched tensile strengths and, in some instances, the work of fracture of notched samples. AE monitoring and microstructural studies were carried out in support of the mechanical tests. Under various circumstances, the effects of microstructural damage resulted in independent changes in the notched and un-notched strengths of a composite. The toughness to strength ratio, K_Q/σ_f, either rose or fell, depending on the material and the damaging conditions. However, the results for all of the tests presented still fell within the 90% confidence limits for the K_Q/σ_f relationship previously identified.

Ma et al [114] examined the hydrothermal effects on the fatigue behaviour of the CF/PEEK laminates before and after impact damage. $[0/45/90/-45]_{2s}$ AS-4/PEEK laminated composites were immersed in 80 °C hot water for 45, 90 and 200 days, subjected to falling weight impact with an energy of 8.58 J and then immersed in 80 °C hot water for 45 days. It was found that the tensile strength of AS-4/PEEK laminated composites decreased with the increase of exposure period. The injured AS-4/PEEK composites were subjected to a static load and tension-tension fatigue at various stress amplitudes. The effect of stress amplitude on the fatigue life was studied. The experimental fatigue life for different stress amplitudes were estimated by the median rank order statistic cumulative distribution function. Subsequently, the fitting curves for estimated data were analyzed by the Weibull distribution function. The S-N curves for a series of cyclic loads at various survival probabilities were presented. The damage behaviour of composites subjected to fatigue was also studied by SEM. The fatigue life of immersed specimens was shorter than that of the pristine specimens; overall, the impact damage affected the fatigue life of the composites significantly.

The effects of LVI and cyclic thermal loading on the fatigue behaviour of CF/PEEK quasi-isotropic laminates were examined [115]. The fatigue behaviour of the pristine composites, LVI composites, and LVI and thermally exposed composites were investigated. Cyclic thermal loading was performed in the temperature range between 60 and 260 °C. The residual tensile strength was measured to understand the influence of LVI on the retention of tensile strength. Fatigue testing was performed at a stress ratio of 0.1, with a frequency of 3 Hz. The Weibull distribution function was employed to model the ultimate tensile strength and fatigue life. S–N curves were plotted and the influence of thermal cycling and LVI on the fatigue performance of the CF/PEEK laminates was investigated. The stiffness variation during fatigue testing was monitored and the differences in stiffness reduction for three test conditions were compared. C-scan was used to investigate the damage zone under different LVIs and to understand damage propagation during fatigue testing. Moreover, SEM was used to examine the fracture morphologies of CF/PEEK composites for both tensile and fatigue failure.

The effects of laminate thickness on the tension-compression fatigue behaviour of a quasi-isotropic CF/epoxy composites subjected to LVI were investigated [116]. The penetration impact properties and the residual tensile strength under various LVIs were experimentally defined. The initial slope during penetration impact revealed that the thicker laminate had a higher stiffness than the thinner one. Moreover, the results showed that the total energy required to penetrate the composite was not proportional to the thickness. Static tensile strength measurements and tension-compression fatigue tests at various stress levels were also carried out on the impacted and un-impacted laminates. The median-rank method and Weibull distribution function were applied to predict the failure probability of the composites under given applied loading and fatigue cycles. The relationship between stress levels and fatigue life (S-N curve) was established and reductions in fatigue life of the impacted laminates at various stress levels were compared. The different slopes of the S-N curves implied that different fracture mechanisms were activated, depending on the laminate configuration. The S-N curves of the composites with different thicknesses were also compared. Ultrasonic C-scanning was used to examine the damage zones in composites subjected to various impact energies and fatigue loading.

Plain-weave CFRPs were subjected to drop-weight impact at energies of up to six joules [117]. Tensile and compressive residual-strength tests were performed over the range of impact energies and two energy levels were selected for evaluation of the fatigue performance. Fatigue load amplitudes were chosen to be fractions of the static strengths. The behaviour of impact-damaged specimens was investigated under tension-tension and tension-compression fatigue loads. The sequence of loading (impact/cycle or cycle/impact) was reversed to study its effect on the strength or the life of the composite materials. The experimental data on impact tests were compared with the corresponding results obtained by using a theoretical model. Cyclic loading followed by impact gave better performance than impact damage followed by cyclic loading, and tension-tension performance was superior to tension-compression cyclic loading.

An investigation was undertaken to determine the effects of hail damage on the fatigue strength of a graphite/epoxy composite laminate [118]. Sixteen-ply coupons were subjected to simulated hail impact by ice-balls of 25.4 or 38.1 mm in diameter. Upon impact, the 25.4 mm diameter ice-ball possessed 7.1 J of kinetic energy, while the 38.1 mm diameter ice-ball possessed a kinetic energy of 27.4 J. For further comparison, additional coupons were impacted with a 12.7 mm diameter aluminium sphere with either 7.1 or 27.4 J of kinetic energy. Inspection revealed that the ice-ball impact events did not cause internal damage, while each aluminium sphere impact caused delaminations within each coupon. After being impacted and inspected, each coupon was subjected to constant amplitude tension-tension fatigue. It was determined that neither the 25.4 mm diameter ice-ball impact nor the 38.1 mm diameter ice-ball impact affected the fatigue performance of this particular laminate. The 7.1 J aluminium sphere impact did not affect the laminate performance either, while the 27.4 J aluminium impact did deteriorate the fatigue properties of the material.

Chotard et al [119] reported detailed results on the impact damage and residual mechanical behaviour of pultruded GF/polyester structures. The impact aspects studied included damage analysis and the impact behaviour of distinct geometries. The influence of test parameters such as impact velocity, impactor mass and impactor size on the type of damage observed were emphasised. Four-point flexural tests were performed on box beams and U sections in order to determine the post-impact residual performance of these beams under static and fatigue loading. Several types of failure mode were observed on damaged and undamaged specimens. The residual performance strongly correlated with the impact testing configuration. Moreover, the residual performance was influenced by the profile geometry (in fatigue bending tests) and especially by the geometrical characteristic of the structures (opened for U sections and closed for the box beams).

The main objective of Margueres et al [120] was to monitor the stiffness reduction and damage evolution of a composite GFRP obtained by resin transfer moulding (RTM). The study was based on mechanical tests such as tension, torsion, impact, fatigue and post-impact-fatigue. Ultrasonic tests, the analysis of the amplitude of AE signals, SEM observations and a numerical analysis completed their study. The material translucence allowed for a direct visual analysis. The results highlighted the reliability of ultrasonic measurements to follow the damage evolution in a composite. This approach was proposed for the definition of residual properties of composite materials damaged during service.

Chotard et al [121] presented an experimental investigation on the residual mechanical behaviour of patch-repaired composite pultruded structures initially submitted to LVI loading. They reported detailed results about static, fatigue and impact tests performed on different GF/polyester impact-damaged structures repaired by low-cost manual techniques. All the tests were conducted at room temperature. For all the studied structures, the initial static strength was completely recovered via the repair process. On the other hand, for two types of pultruded structures, the fatigue crack-growth life did not recover its initial values but when compared with the damaged specimen, the lifetime of the repaired structure was significantly enhanced. Residual performances of both undamaged and repaired specimens were influenced by the profile geometry (in fatigue bending tests) and especially by the geometry of the structures. Carefully designed, external scarf patch repairs can recover more than 85% of the undamaged mechanical behaviour, depending on the type of residual applied loading (Figure 21).

Fatigue tests were conducted on impact damaged coupons of T300/914 CFRPs [122]. Two damage levels were used to represent barely visible impact damage (BVID) and visible impact damage (VID). Fatigue tests were conducted at a frequency of 5 Hz and a load ratio of $R= -1$. Progression of fatigue damage was monitored by ultrasonic C-scan, measurement of changes in the coupon modulus and measurement of hysteresis in the impact damaged coupons.

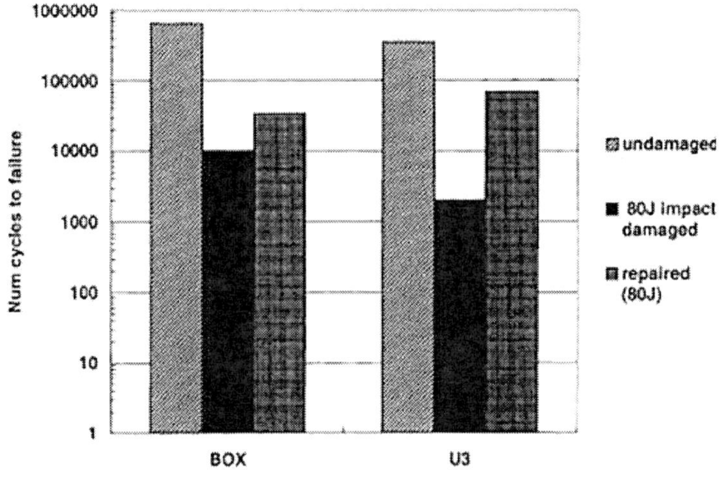

a

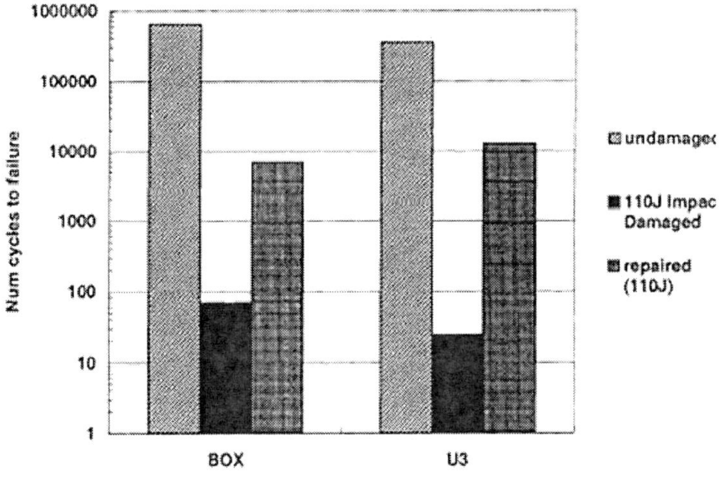

b

Figure 21. Residual fatigue damage for undamaged, impacted and repaired composites: (a) 80 J impacted specimens and (b) 110 J impacted specimens [121].

Horn et al [123] proposed a new experimental method for quantifying impact damage and estimating the remaining fatigue lifetime of impact damaged polymer matrix composites. The procedure was demonstrated using GF reinforced polyurethane produced by injection moulding and structural reaction injection moulding. Thermoelastic stress analysis (TSA) was used to quantify the stress concentration associated with impact-damage in test samples for each composite. Following impact and TSA imaging, the samples were fatigued to failure over a range of stress amplitudes. The TSA-derived stress concentration factors were used to determine a modified stress amplitude that integrated the impact-fatigue data onto a master stress-life curve. This approach provided a quantitative measure of impact damage and a practical methodology for estimating the residual fatigue lifetime of impact damaged composites.

Impact-damaged CF/epoxy composite laminates were tested in tension-compression fatigue under constant-amplitude loading with emphasis on studying mechanisms leading to delamination growth and fatigue failure [124]. The shapes of the buckles were measured with an optical whole-field measurement technique, while the extension and depth of delaminations was imaged several times during testing by ultrasonic C-scanning. The delamination growth occurred mainly transversely to the load application direction and buckles observed on the backside usually had the same shape as some of the delaminations. This indicated that buckling which took place during the compressive part of the load cycle instigated delamination growth. In specimens run to the fatigue limit, buckling and delamination growth occurred only in the outer 2 or 3 layers on the backside while a specimen that failed before the fatigue limit buckled globally.

Constant amplitude tension–compression fatigue was performed on two layups of impact damaged CF/epoxy composite laminates [125]. The shape and the amplitude of the buckles were measured using an optical whole field measurement technique. The experimental conditions were chosen to be representative of damage in aircraft structures. The R-value was found to have a small influence on the fatigue life indicating that the compressive part of the load cycle had more importance than the tensile part, as the compressive load caused local buckling around the damage zone. The buckling was inward on the impact side and outward on the backside with larger buckles on the backside. Also the backside buckles showed a larger growth. Fatigue lives were compared with a conservative analytical model.

Puhui et al [126] presented a new method for the determination of the CAI strength of composite laminates. In this method, the impact damage zone was modelled as an equivalent hole. The most outstanding characteristic of the method was that the simplification of the impact damage was based on the compressive failure mechanisms of impacted laminates. Such a simplification was not discussed in the literature before. A technique was established for determining the shape and size of the equivalent hole. The stress distribution around damage was calculated using the complex potential method and the classical lamination theory. A lay-up independent failure criterion was used to predict CAI strength. The predictions of the present approach were compared to test results and were found to be in very good agreement over a wide variety of materials. The study indicated that the width of damage zone was a key factor governing CAI strength. The influence of damage area, impact energy, impactor shape and dimension, etc. on CAI strength was characterized by their effects on the damage width. The study provided a very simple and effective approach for CAI strength prediction compared with previous methods.

Lee et al [127] investigated the relationship between LVI damage under static indentation loading and the residual compressive static and fatigue strength in hybrid composites with nonwoven carbon tissue (NWCT). The hybrid laminates were made by interleaving the NWCT layer in the CFRP layer interfaces. The indentation damage was associated to both delamination area and indentation energies of angle-ply laminates. The delamination area of the hybrid laminates were reduced to about half compared to the area induced in plain CFRP by indentation loading. After the indentation damage, compressive static and fatigue tests were carried out using an anti-buckling guide for the angle-ply laminates. The failure modes of the angle-ply CFRP and hybrid specimens were caused by global buckling under the compressive static loading. However, the failure modes of the angle-ply CFRP and hybrid specimens at $N_f < 10^6$ cycles were caused by shearing of the damaged part under the compression–compression fatigue loading. Compared with CFRP laminates, compressive fatigue lives of the hybrid laminates were significantly extended in all the stress ranges. The NWCT interleaving effect was assessed based on the damage extent in the hybrid specimens after indentation. The compressive static and fatigue failure processes were discussed from the observation of the fracture surface of damaged parts and the appearance of failed specimens when subjected to the compressive loadings after indentation.

The impact properties and the fatigue properties of sandwich composites after impact were addressed [128]. Specifically, the LVI resistance of CF faced sandwich composites and the effect on their fatigue life was studied. Two and four layer face sheet CF sandwich composite samples with foam filled honeycomb core were impacted in a low velocity drop tester with varying masses and velocities of the impactor. After impact, the area of damage was determined using ultrasonic techniques. The fatigue life was then determined and compared to the fatigue life of non-impacted samples in a four point bend test. For all impact energies the impactor completely penetrated the front face of the sample. At 20 and 30 J the impactor also penetrated the back face. The 30 J caused a larger penetration area whereas the 20J caused more delamination of the back face. Two different densities of foam were used for the two layer samples. The two layer high density sample failure mode changed from in shear for the virgin materials to bending for the impacted samples. The penetration from the top face deteriorated the fatigue properties of the two layer samples. The fatigue life of the four layer and two layer low density samples did not appear to be affected by the impact.

The damage tolerance of various lay-ups of thin carbon/epoxy laminates was examined [129] by CAI tests, using a testing device which adapted to the thicknesses of the specimens and did not require tabs or any modification of the specimen geometry. The compression stress state was not modified by the device geometry, as was verified by numerical simulation. With this device, CAI tests were performed for different carbon/epoxy laminate lay-ups (quasi-isotropic, cross-ply and woven) and the values of the residual strength and the normalized residual strength of the laminates were obtained as a function of the impact energy. The woven laminate was found to possess the highest residual strength for all impact energies, and the quasi-isotropic laminate the least normalized strength loss with increasing impact energy.

Petit et al [130] presented an experimental study of impact and CAI tests performed on composite laminate covered with a cork thermal shield (TS) intended for launchers' fairings. Drop weight impact tests were performed on composite laminate sheets with and without TS in order to study its effect on the impact damage. The results showed the TS was a good mechanical protection towards impact as well as a good impact revealing material. Nevertheless, totally different damage morphology was obtained during the impact test with or without TS, and in particular at high impact energy, the delaminated area was larger with TS. CAI tests were also performed in order to evaluate the TS effect on the residual strength. The TS increased the residual strength for the same impact energy, but at the same time, it caused a decrease in the residual strength for energies where delamination was not visible. As was postulated, during the impact tests with TS, invisible fibres' breakages took place before delamination was visible, which was not the case for the unshielded sheets.

Xiong et al [131] studied the static CAI and fatigue properties and the failure mechanisms of CF reinforced composite laminates with different lay-ups in order to optimize the stacking sequence. Ten different lay-ups were studied and the results for static strength, CAI and fatigue residual strength under different fatigue stress amplitudes were evaluated and compared to each other. The results of this study provided an insight into fatigue damage development in composites and constituted a fundamental basis for the development of a strain-based residual strength model. A new formulation was presented for a fatigue-driven residual strength surface based on controlling fatigue strain. The determination of the parametric formulae of the residual strength surface was established effectively and easily for the experimental data. The application of the model to test data, demonstrated its practical and effective use. The fatigue-driven residual strength could be obtained realistically from experimental data using the new formulae.

The LVI and tension–tension fatigue behaviour of glass fibre reinforced polyester resin composites were investigated by Yuanjian and Isaak [132]. Two fibre geometries, namely $[\pm 45°]_4$ and $[0/90°]_{2s}$ stitch bonded GF were studied. The results revealed that LVI could seriously impair the tensile properties of $[\pm 45°]_4$ composites. For the $[0/90°]_{2s}$ GF composites, a critical impact energy was found. Below this energy level the tensile properties were hardly affected by the impact but above it the tensile properties deteriorated with increasing energy. LVI also reduced the fatigue lives of the composites and this reduction was related to the degradation in tensile strength. By normalizing the fatigue stress against the post-impact residual tensile strength of the composite, it was found that for each fibre geometry a single *S–N* curve could be drawn, which encompassed both undamaged and impacted samples. This implied that fatigue lifetimes of impact damaged composites could be predicted from measurement of the residual tensile strength of impacted specimens and the *S–N* curve of undamaged samples.

Saito and Kimpara [133] evaluated the damage evolution behaviour considering the effect of the textile structure and water absorption. Damage observation was conducted using both non-destructive and direct observation methods. Candidate textile reinforcements were T300-3k plain woven fabric (PW) and T700S-12k multi-axial knitted fabric (MA) (Figure 22). The effect of water absorption on the performances of CAI and Post-Impact-fatigue (PIF) were small in PW CFRP laminates. Conversely, PIF properties of water-absorbed MA CFRP laminates drastically decreased. CAI strength was not affected by water absorption. PIF performance of dry MA CFRP was fairly higher than that of the other configurations. Using microscopic observation, some evidence of interfacial deterioration caused by water absorption was confirmed for both PW and MA CFRP laminates.

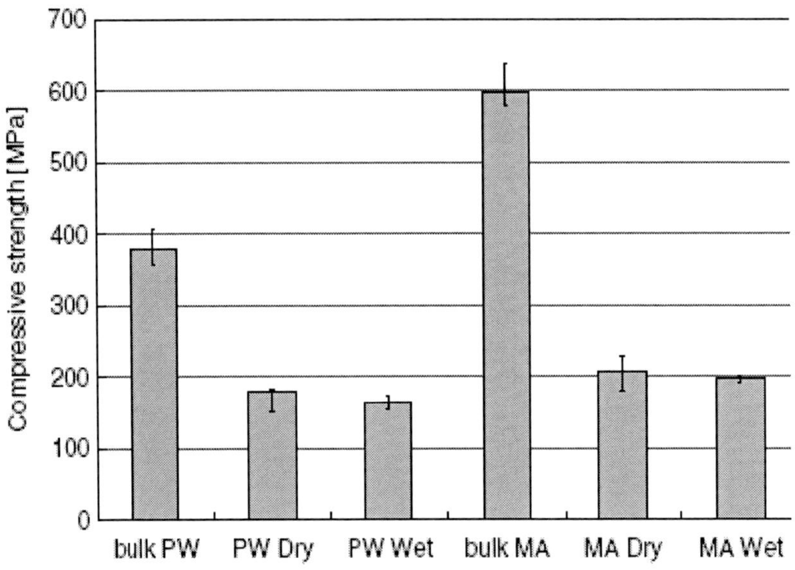

Figure 22. Compressive strength after impact for plain woven and multi axial knitted CFRP under dry and wet condition [133].

Minak et al [134] dealt with the relation between damage and tension–tension fatigue residual strength (FRS) in a quasi-isotropic CF reinforced epoxy resin laminate. The work was organized in two phases: during the first one, composite laminates were damaged by means of an out-of-plane quasi-static load that was supposed to simulate LVI; in the second phase, fatigue tests were performed on damaged and undamaged specimens obtained from the original composite laminates. During the quasi-static transverse loading phase, damage progression was monitored by AE. The measurement of the strain energy accumulated in the specimens and of the acoustic energy released by fracture events made it possible to estimate the amount of induced damage and evaluate the quasi-static residual tensile strength of the specimens. A probabilistic failure analysis of the fatigue data, reduced by the relative residual strength values, led to the correlation of the FRS of damaged specimens with the fatigue strength of undamaged ones.

Natural Composites

Jute/polyester woven laminates have been impacted to energies up to 20 J with an impact velocity of 2 m/s using a drop-weight impact tower [135]. The diameter of the hemispherical impactor was 12.7 mm. Some of the laminates were further subjected to post-impact tensile tests to failure. The aim of these tests was to measure the degradation of their mechanical performances with increasing impact energy. Tensile failure occurred by macroscopic matrix cracking under the point of impact. The interlaminar fracture toughness of this material was sufficiently high and failure did not involve delamination. Fatigue loading with a maximum applied stress of 13 MPa i.e. around 25% of the ultimate tensile stress of the laminate was also performed. The fatigue ratio was 0.5 and the loading frequency used was 10 Hz. The damage produced by impact and the subsequent fatigue loading was observed using an optical microscope.

Post – Impact Static Behaviour Modelling

A semi-empirical analysis on residual compressive strength (RCS) of CF/epoxy woven composite laminate was developed [136] which included the damage effects caused by impact and hydrothermal cycling. Impact damage was modelled as a soft inclusion with an exponentially decreasing stiffness, which was further reduced due to hydrothermal cycling. A complex variable method was used to determine the in-plane stress distribution near the impact induced damage and a point stress failure criterion was then used to predict the failure load. Based on the semi-empirical model, the RCS could be related to damage width, damage intensity, undamaged strength and a degradation factor due to hydrothermal cycling. The results from the analysis coincided reasonably well with the experimental data for plain-woven fabric laminates.

Nilsson et al [137] presented a combined numerical and experimental study of slender composite panels loaded in compression with artificial delaminations at two different depths. The study was motivated by FEA where this change in delamination depth induced a transition in the direction of delamination growth along with a change in the basic fracture models and stability. Tests were then carried out to verify the transition in delamination growth. The predicted transitions were to a large extent verified experimentally. The paper gave an outline of the computational model, which included contact between delaminated members, calculation of energy release rate with fracture mode separation by an approximate as well as a reliable method for general lay-ups, and a moving mesh scheme to account for delamination growth. The experimental work included monitoring of delamination growth by C-scan and acoustic emission along with a detailed fractographic study. The correlation between experimental observations and computer results were discussed in detail.

Previous solutions to the low mass impact problem were based on laborious numerical approaches. Olsson in his paper [138] presented a closed form solution based on three asymptotic cases with deformation purely by indentation, bending or shear. Solutions were presented for a Hertzian contact law, suitable for monolithic plates, and for a linear contact law, suitable for sandwich panels. A good agreement was shown between "exact" numerical solutions and experimental results. The solution was also combined with an existing quasi-static delamination threshold load criterion. The comparison with data published in the literature indicated that the criterion also may be applicable to small mass impacts.

The effect of delaminations in a post-buckling stiffened structure manufactured from laminated composite materials was modelled [139]. Driven by aircraft certification requirements, the emphasis of the technique was towards establishing whether delamination growth would initiate under given loading conditions. A geometric non-linear FEA was used to calculate the SERR around the circumference of a circular delamination using the virtual crack closure technique. In order to deal with the complex structural response in a computationally efficient manner, the structure was modelled using plate elements with two layers of plate elements used in the delaminated region. The effect of delamination size on the post-buckling strength panels was shown to be a complex phenomenon in which trends were difficult to predict. Large delaminations could significantly affect the global and sub-laminate buckling modes and therefore be less critical than smaller delaminations. It was concluded that the method could accurately predict the load and location at which delamination growth would initiate with the provision of reliable critical Strain Energy release Rate (SERR) data.

Tay and Shen [140] presented the results of a buckling and post-buckling analysis and modelling of embedded delaminations, and the prediction of delamination growth in laminated composites with consideration for residual thermal stresses. The distribution of the local SERR along the delamination front was obtained via the virtual crack closure technique applied to three-dimensional (3D) FEA models of circular delaminations embedded in woven and nonwoven composite laminates. In each case, the delamination was embedded at a different depth through the thickness of the laminates. The fibre orientation of the plies bounding the delamination influenced significantly the distribution of the local SERR. There was qualitative agreement between the predicted directions of delamination growth and C-scan results of the embedded delaminations. It was found that residual thermal stresses affected significantly the onset of buckling of the delaminated sub-laminate, but had negligible influence on the distribution of the local SERR in the post-buckled regime. Furthermore, the effect of residual thermal stresses was more pronounced for delaminations that were closer to the surface. A method for the prediction of delamination areas and directions using fatigue growth criteria derived from test coupon data was also presented. It was found that growth criteria based on components of the SERR predicted the rate of delamination growth much better than those based on the total SERR.

Riccio et al [141] performed a crack growth analysis on composite panels containing embedded delaminations using a geometrically non linear FEA code based on the total Langragian formulation. The code was improved with an effective virtual crack closure technique to evaluate SERR and with a penalty method for contact forces. The validation of the proposed tool was performed with experimental and numerical data available in the literature for double cantilever beam (DCB) specimens. Finally, the influence of the geometrical parameters of the delamination (size and location along the thickness) on the SERR distribution and delamination growth stability in composite panels under compression was analyzed.

In a following article [142], Riccio and Gigliotti proposed a novel numerical approach for the delamination growth simulation in composite panels under compressive load. This approach, suitable for preliminary design and optimization purposes, was able to simulate the delamination propagation by means of a limited number of linear analyses. It was based on the determination of delamination buckling and on the evaluation of the energy released during the delamination propagation by means of eigenvalue and linear static analyses. The proposed approach was implemented into the FEA code ANSYS® and applied to composite panels with circular embedded and rectangular through-the-width delaminations under compressive load. A first validation was carried out by comparing the results of delamination growth load and SERR distributions along the delamination front to two-dimensional and three-dimensional nonlinear results taken from literature.

A numerical simulation of the structural behaviour of delaminated composite plates under compressive load was developed and presented by Riccio and Pietropaoli [143]. In this article the effect of fibre – matrix failure on the buckling behaviour of a delaminated structure up to final failure was investigated in detail. A geometrically non linear FEA approach was adopted for elastic instability simulation and the modified virtual crack closure technique (MVCCT) was used for the SERR evaluation at the delamination front. A fracture mechanics progressive damage approach was introduced for the simulation of fibre – matrix damage onset and evolution. Comparisons of numerical results with existing experimental data on composite panels with an embedded circular delamination under compressive load were presented for preliminary validation purposes. Furthermore, the influence of the different failure mechanisms on the compressive behaviour of delaminated composite plates was assessed, by comparing numerical results obtained with models of different degrees of complexity. Finally, results of a limited sensitivity analysis were presented, pointing out the influence of some key parameters of the adopted numerical approach on the accuracy of numerical simulations.

Surface free energy has been often treated as a scalar constant without considering its dependence on propagation direction. It is desirable, however, to investigate how surface free energy or fracture toughness of delamination in a single interface varies with both the local mismatch angle of fibre directions and the direction of crack propagation in polymeric laminate composites. As a materials constant, fracture toughness is effectively used for various mechanical analyses of fibre-reinforced composites as well as conventional materials. Kim and Mayer [144] investigated quantitatively and qualitatively the dependence of delamination fracture toughness on mismatch angle and crack propagation direction in laminated structures. AS4-Carbon/Epoxy prepregs were used for fabricating test specimens, and 50 different mismatch angles of fibre direction were applied on the delaminated interface of laminates. Fracture toughness was measured using the Mixed-mode Bending (MMB) test. This test method imposed both mode-I and mode-II fracture, and the mixed-mode ratio (G_{II}/G) can be controlled. The mixed-mode ratios used here were 20, 35, 50, 65, and 80%. The crack path and the delamination fracture toughness were observed and calculated for specimens, and the dependence on fracture toughness was shown to be related to the mismatch angle of ply fibres at the delaminated interface. The relationship between mismatch angle and delamination fracture toughness was newly revealed and discussed for various angles (Figure 23). These results could be usefully applied to various fracture mechanics analyses in fibre-reinforced laminated composites.

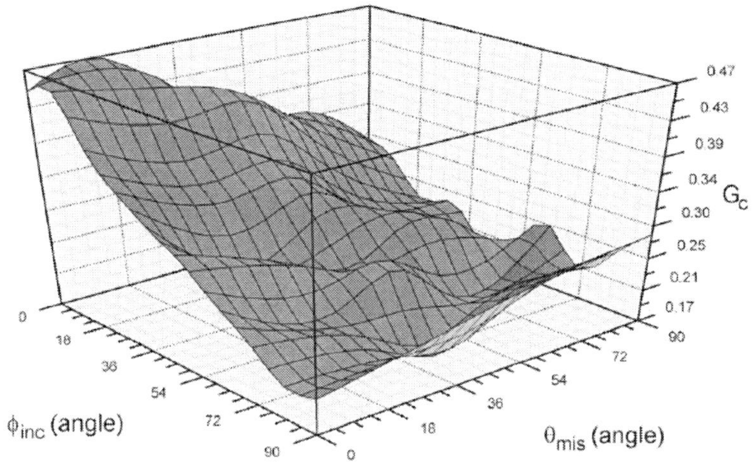

Figure 23. Fracture toughness (G_c) as functions of incline angle and mismatch angle at mixed mode ration G_{II}/G =80% [144].

Attia and Kinloch [145] attempted to predict growth of impact damage in CFRP skin/stringer structures subjected to cyclic-fatigue loading. They proposed a method that combined an experimental relationship between the SERR, *G*, obtained from direct measurements and the number of fatigue cycles N_g with a linear FEA of the composite structure under cyclic-fatigue loading.

Kang and Kim investigated the fatigue behaviour and variability of fatigue life in carbon/epoxy laminates with impact-induced damage under tensile [146] and 2-stage tensile [147] fatigue loading. To describe the fatigue behaviour of the impacted laminates, the strength reduction concept was introduced, based on the Broutman's model for 2- and 3- stage loading, respectively. By using this concept combined with the expected fatigue behaviour of the un-impacted laminates, a fatigue life prediction model for the impacted laminates was derived and then verified through the fatigue tests on the impacted laminates under constant amplitude loading. In order to account for the probabilistic nature of the phenomenon, a statistical model was developed by introducing the random variable Z: The model described well the distribution of fatigue life in laminates. As was reported, the variation of fatigue life decreased rapidly as the applied impact energy increased.

A test method, with specimen design similar to that proposed by O'Brien et al. [63, 148, 149 150] but in 3-point bending, was proposed [151] to measure the critical SERR (G_c) for delamination in fibre-reinforced polymers (FRP) under out-of-plane or transverse loading. Unlike end-notch-flexure (ENF) or double-cantilever-beam (DCB) tests that are currently used, the proposed method did not require an insert film or a pre-crack to initiate delamination. Instead, the delamination was initiated from a FRP layer of the same composition but with different fibre orientation. For this study, the FRP layer orientation was 90°, i.e. one 90° layer in the mid-thickness and the rest 0° layers. Under 3-point bending, matrix shear cracking was firstly initiated in the 90° layer, mimicking the mechanism for delamination initiation in the FRP when subjected to transverse loading. The matrix shear cracking led to delamination in the adjacent interlaminar regions and the delamination area was measured after the test. Calculation of G_c was based on “the area method”, that is, dividing the total energy loss by the delamination area. The G_c was compared to that from plate specimens of the same composition and fibre lay-up but subjected to transverse point loading. The comparison indicated that very similar G_c values were acquired from the two types of testing, even though unstable crack growth occurred under 3-point bending and stable crack growth under transverse point loading. Therefore, the proposed method could potentially quantify delamination resistance of FRP under the transverse loading conditions.

Andrews and Massabò [152] studied the static and dynamic response of singly and multiply delaminated plates subjected to cylindrical bending using models based on Timoshenko‘s shear deformable beam theory. The static interaction of multiple delaminations and its effects on fracture parameters and macro-structural behaviour was investigated. Interaction effects led to phenomena of amplification and shielding of the SERR and modifications of the mode ratio when compared to systems with a single delamination. The behaviour was controlled by the through thickness spacing of the delaminations. The dynamic delamination fracture of a double cantilever beam was studied and the capability of the proposed model to follow the different phases of delamination growth, including arrest, was highlighted [153]. The effect of loading pulse duration on stationary delamination systems was investigated and the response was described by SERR shock spectra to highlight phenomena of dynamic amplification. The results of the proposed models were compared with two-dimensional FEA solutions and showed excellent agreement.

A computational model for investigating the local extent and the through-the-thickness position of the damage induced by LVIs on laminated composites was presented [154]. The model was based on a refined 3D zig-zag approach. It fulfilled the interlaminar transverse shear and normal stress contact conditions at the layer interfaces, those on the transverse normal stress gradient and the boundary conditions at the upper and lower bounding faces, as required by the theory of elasticity. Its functional degrees of freedom are the three components of the elastic displacement and the two shear rotations, like in widespread smeared laminate models. Several classical models were particularized from it and confronted together to assess whether a sophisticated structural model could have practical advantages for the study of impacts, since this is a still open question. In order to improve the structural modelling with an affordable computational effort, the contributions of the present model were incorporated updating the strain energy of a C^0 parent eight-node plate element based on the First Order Shear Deformation Plate theory and through a post-processing procedure based on spline interpolation of the involved quantities. The impact load was computed by the Hertz's contact law, integrating the equilibrium equations by the Newmark's algorithm. The Galerkin's method was used to obtain these equations, instead of using FEA discretization, because accurate results were obtained at lower costs. Fibre, matrix and delamination failures were predicted using two different strength-based criteria for each single mode. Since their predictions could greatly diverge, a comparison provided an estimation of the variation range so as to choose the appropriate one for each failure mode. According to the ply-discount theory, the degradation of properties after failure was simulated by reducing the elastic properties of the failed plies, in conformity with the failure mode. Different models with a different representation of displacements and interlaminar stresses were shown to provide different results in terms of displacements, contact force and damage. The application to a laminated stiffened panel indicated that the discrete-layer effects played an important role in the simulation of impacts, even when the laminates were not thick, since high-order dynamic effects were involved. The structural model, the contact force simulation, and the failure criteria simulated the impact induced damage, extent and position across the thickness with a satisfactory accuracy. This was verified by comparison with ultrasonic detection.

Impact-induced damage in fibre-reinforced laminated composite plates was characterized by Watkins et al [155]. An instrumented impact tower was used to carry out LVIs on thirteen clamped glass/epoxy composite plates. A range of impact energies was experimentally investigated by progressively varying impactor masses (holding the impact height constant) and varying impact heights (holding the impactor mass constant). The in-plane strain profiles as measured by poly-vinylidene fluoride (PVDF) piezoelectric sensors were shown to indicate damage initiation and to correlate to impact energy. Plate damage included matrix cracking, fibre breakage, and delamination. Electronic shearography validated the existence of the impact damage and demonstrated an actual damage area larger than visible indications (Figure 24). The strain profiles that were associated with damage were replicated using an in-house built FEA code. Using these simulated strain signatures and the shearography data, a back-propagation artificial neural network (ANN) was employed to detect and classify the type and severity of damage. This result was an extension of a prior work that showed that a strain-based neural network could characterize impact energy and contact forces for non-damaging impacts [156].

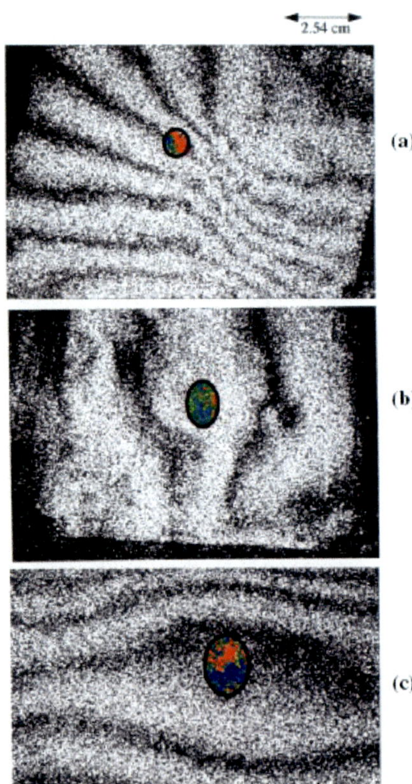

Figure 24. Shearography images of impacted plated (a) 5.9 J, (b) 7.8 J and (c) 12.9 J. The damaged areas are within the marked ellipses [155].

An analytical model for predicting the compressive fatigue limit strain of composite laminates, which contain barely visible impact damage was presented in [157]. The model represented the complex damage morphology as a single, circular delamination, and calculated the strain at which thin-film buckling of the circular region of delaminated plies occurred. The fatigue limit strain was defined as the strain at which the SERR for a thin post-buckled strip of the delaminated plies was equal to the critical Mode I value (G_{IC}) for the resin. The model predicted a "critical" depth at which propagation of damage during fatigue was likely to occur (Figure 25). Results obtained using the model were compared with two sets of experimental results, and showed agreement of fatigue limit strain to within 4% of the experimental value.

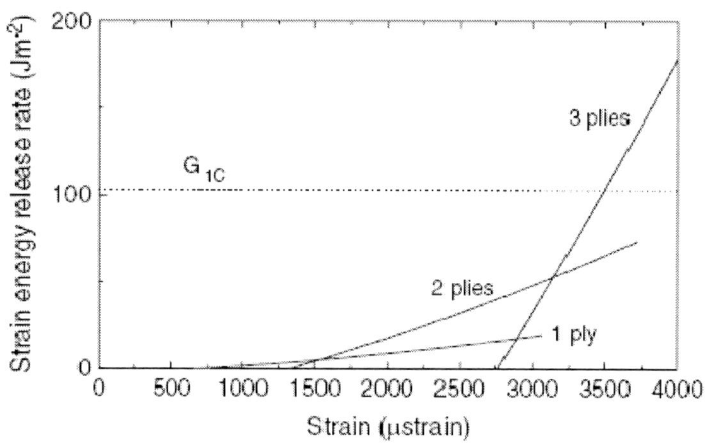

Figure 25. Strain Energy release rate for delamination at 1, 2, and 3 ply depths in [45, -45, 0, 90]$_{2s}$ CFRP laminates [157].

The influence of compressive stress on mode II damage evolution has been investigated [158] based on the cut-ply and dropped-ply experiments. An interfacial failure model with modified failure initiation and propagation criteria was proposed in order to take into account the effect of compression on matrix shear strength and mode II critical fracture energy, G_{IIC},. The model used one independently determined parameter to relate the compression to the increase in interlaminar shear strength and G_{IIC}. Using this failure criterion, two types of cut-ply models and two types of dropped-ply models, using the same input parameter produced excellent correlation with experimental delamination stresses. As a validation case, the single-lap model was run with the above criterion and the same input parameters. The model correlated well with the experimental failure stress.

The dynamic compressive strength of quasi-isotropic CF/epoxy composite was investigated experimentally and simulated numerically [159]. In-plane compression tests at strain rates around 400/s of quasi-isotropic laminates were performed using the Split Hopkinson Pressure Bar (SHPB). The dynamic strength of quasi-isotropic laminates exhibited a considerable increase when compared to the static values. The FEA model used ABAQUS™ three-dimensional solid elements C3D8I with 8 nodes and a user-defined interface with 8 nodes finite elements. These interface elements interrelated the three-dimensional solid elements which modelled the composite layers and included a cohesive damage model which allowed for the simulation of delamination initiation and propagation. The proposed model assumed that the phenomenon of failure under these conditions was mainly dictated by interface delamination. This was supported by experimental tests which showed that all quasi-isotropic laminates split into several almost intact sub-laminates. The model compared very well with experimental results, confirming the formulated hypothesis that the internal layer damage did not markedly contribute to the quasi-isotropic laminate failure.

Craven et al [160] examine the effect of impact damage on local tensile stiffness of multi-directional tape laminates. A FEA model was used for parametric studies on the effect of various patterns of delaminations and cracks across the fibres. Regular and random crack patterns were considered for 8 and 16 ply quasi-isotropic laminates. The effect on local stiffness was evaluated by a recently developed inverse numerical approach. It was found that although fibre cracks control the loss of stiffness, delaminations were instrumental in extending the zone influenced by these cracks. Fractographic observations were used to generate a detailed model of the damage in a real laminate. A close agreement between the predicted and measured strain fields and the corresponding stiffness distributions calculated using the inverse method was found.

Chapter 6

DELAMINATION GROWTH PREDICTION (CYCLIC)

The most up-to-date research, in the mechanics of delaminations and related crack-like defects in laminate and fibre composites, was discussed in [161] and was updated in [162]. Both internal and near-surface delaminations were considered. In the latter case, local buckling of delaminations and the interaction between buckling, damage accumulation, crack growth and global buckling were considered. The problem of the evaluation of the residual load-carrying capacity of delaminated structural components was discussed, including the assessment of the fracture toughness with respect to impact loading.

Quasi-isotropic carbon/epoxy laminates with holes, polymer plugs and cut fibres were studied experimentally and analytically [163]. Strain fields were measured using digital speckle photography. The results were used to validate an inverse method, where elastic properties of inclusions were determined by matching computed and measured displacements. The tensile and compressive strength were measured and the applicability of notch failure criteria to soft inclusions was examined. Available closed form solutions agreed well with measured strains. The elastic properties predicted by the inverse method were in fairly good agreement with data from coupon tests although predictions were sensitive to measuring errors. The laminate toughness in compression was higher than in tension, as expected from the different failure mechanisms such as fibre kinking. Laminates with inclusions were tougher than laminates with holes, which suggested that inclusions restrain in-plane fibre kinking. Higher toughness was reflected in larger characteristic lengths.

Fatigue tests of a hat-shape stringer stiffened panel were conducted [164]. The panel was a typical part of upper skin of a lightweight composite wing using a new production technology of stitching, co-bonding and RTM method. Impact damage was induced on the skin/stringer co-bonded part and on the typical skin part of the test panel by a drop-weight impact machine. The test protocol consisted of two phases. The first phase involved fatigue tests to verify durability of the structure with barely visible impact damage. The second phase was flaw growth tests for evaluation of visible impact damage growth to estimate inspection intervals. The Mini-TWIST (shortened version of The *T*ransport *WI*ng *ST*andard load program) spectrum loading was used for both tests. NDI was carried out by pulsed thermography during the test to observe damage propagation. Finally, static load was applied up to the design limit load to verify the residual strength after all the spectrum loading tests.

A composite laminated plate containing an internal delamination and subject to compressive loads was considered [165]. The plate was modelled as two sub-laminates partly bonded together by an elastic interface. The interface in its turn was represented by a continuous distribution of linear elastic springs acting in both the normal and tangential directions with respect to the interface plane. Based on an explicit solution, already presented in previous works by the authors, the expressions for the I and II (opening and sliding) modal contributions of the potential SERR, G, and the mode-mixity angle, ψ, were determined. In the reviewed paper, the model was further developed to describe the mechanical response of delaminated plates under fatigue loads. A mode-dependent fatigue growth law was applied, which accounted for the simultaneous presence of both opening and sliding crack propagation modes. The results could be used to easily predict the number of cycles needed for a delamination to extend to a given length or for complete failure, for any value of the minimum and maximum applied load.

Natural Composites

A series of experiments was carried out to characterize the residual tensile and fatigue after impact properties of non-woven hemp fibre mat reinforced polyester [166]. Additionally, the degradation of tensile modulus during fatigue cycling was studied and related to damage accumulation. For comparison purposes, ±45° GF reinforced polyester samples were also subjected to similar tests. It was necessary to apply a relatively high pressure to the hemp composite during the curing stage in order to ensure a high enough fibre fraction and provide a significant reinforcing effect. With similar fibre weight fractions, the hemp and glass reinforced materials exhibited similar static tensile properties and fatigue lifetimes. Although the slightly steeper *S–N* curve of the hemp based material indicated a higher rate of reduction in fatigue strength with increasing cycles, it remained above the *S–N* curve for the glass based material. It was thus concluded that the hemp-based material was able to withstand slightly higher cyclic stress levels for the same number of cycles. The major difference in mechanical performance was the poorer resistance of the hemp based composite to impact. Additionally, the hemp based material failed in a much more brittle manner, without any visible signs of damage, such as the matrix cracking that was observed in the GF based composite. It was found that, if the fatigue lifetime data of impact damaged samples were normalized against the post-impact residual tensile strength, all data points lay close to a common *S–N* curve. This implied that residual fatigue lifetimes of damaged samples could be predicted from their residual strength and the *S–N* curve of the undamaged material.

Models of Prediction

Katerelos et al [16] studied the fatigue behaviour of composite panels subjected to LVI. Impacted specimens were tested under compression–compression fatigue. A delamination propagation model based on the derivation of the SERR was used. The shape of the delamination was experimentally monitored by C-scan imaging and it was transferred to a polar diagram, where the load direction was at 90°. The stress distribution around the initially induced delamination was derived analytically, based on the assumption that both the initially delaminated area and the delamination growth after certain amount of fatigue cycles was simulated by an ellipse with semi-axes defined from the original delamination shape. The definition of the ellipse simulating a C-scan detected delaminated area was presented (Figure 26).

Stress distribution functions formulated according to the Lekhnitskii – Savin theory were derived. The orientation and aspect ratio of the ellipse were used to calculate the corresponding strain energy-release rates, which were subsequently used to predict the direction of delamination propagation. A flowchart representation of the algorithm for the calculation of the SERR around the ellipse was outlined (Figure 27).

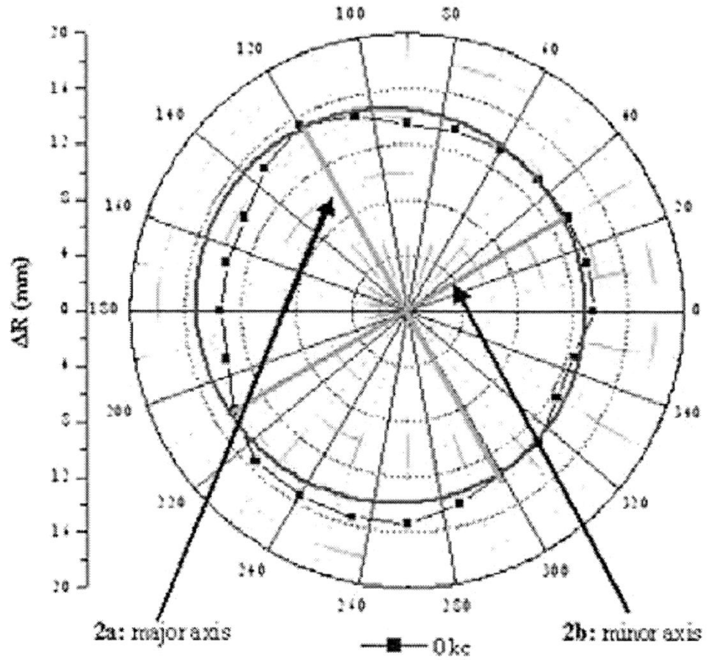

Figure 26. Definition of an ellipse to simulate a C-scan detected delamination.

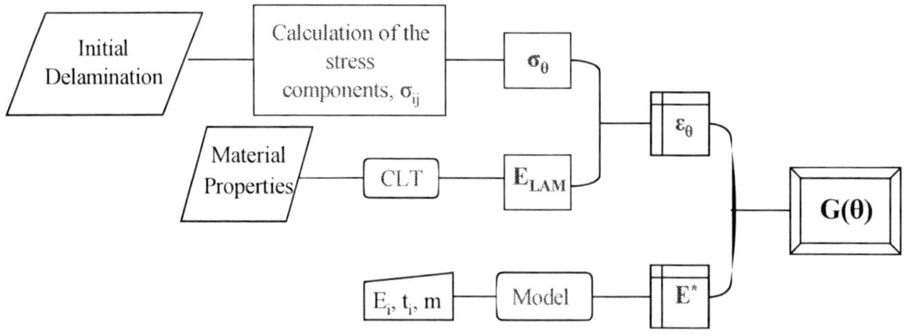

Figure 27. Calculation of the SERR distribution. Flow-chart representation of the algorithm.

The reduced stiffness properties (E^*) of the delaminated plate were derived using the model proposed by O'Brien [63]. Thus the direction of delamination growth could be easily predicted.

In order to quantify the delamination area growth, a power law similar to the Griffith's law was used. The constants of the power law were derived from experimental data. By the application of this law to impacted and subsequently fatigued laminates, with different initial delamination areas resulting from different impact energies, it was proved that the delamination growth rate with respect to the fatigue cycles was independent from the impact energy and, therefore, from the initial delaminated area. The delamination growth rate was predicted as a function of SERR (Figure 28).

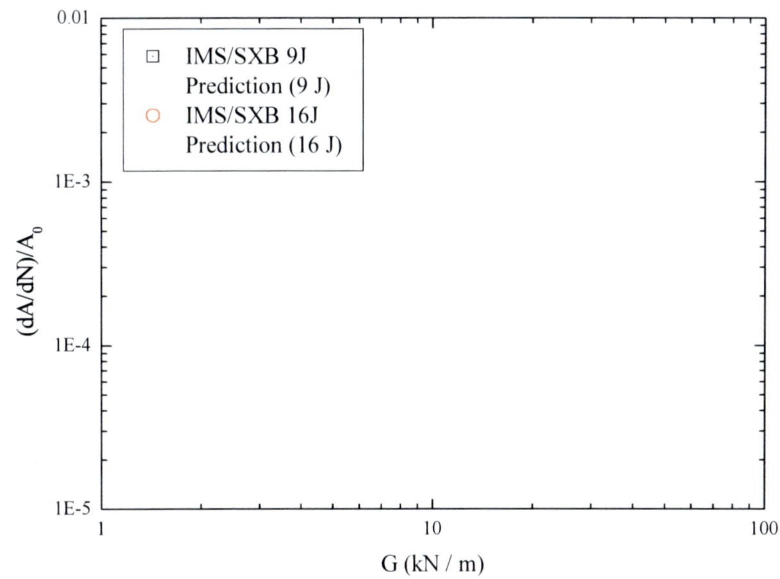

Figure 28. Prediction of the normalised delamination growth vs. the maximum SERR.

Models to predict the split growth in notched AS4/3501-6 graphite/epoxy quasi-isotropic laminates under tension-dominated fatigue were presented [167]. First, a power law model and an artificial neural network (ANN) model were developed to describe the split growth under constant-amplitude fatigue. They were then applied in conjunction with a linear damage growth model to predict the split growth under spectrum fatigue. The ANN model was found to work better than the power law model as a predictive tool for split growth.

A multi-layer perception (MLP) network using error back propagation algorithm was employed by Garg et al [168] in order to estimate the damage parameters from broad-band spectral data as a diagnostic signal. Various existing models of damage in laminated composite and the resulting stiffness degradation were discussed. Degradation of ply properties as a function of transverse matrix cracks in cross-ply, splitting in longitudinal ply, and evolution of consecutive stages of damage, such as delaminations and fibre fracture was considered to be one of the damage model parameters. The stiffness degradation factor and the location and size of the damaged zone in a laminated composite beam were considered as damage model parameters. Fourier spectral data, which were typical for most of the diagnostic wave measurements, were used as input to the neural network. Training of the neural network involved many data sets difficult to generate using experiments, and therefore a spectral finite element model (SFEM) with embedded degraded zone for the laminated composite beam was developed. A numerical simulation using this model was carried out, in order to estimate the temporal signals that were likely to be measured. Analytical studies on the performance of the neural network were presented based on numerically simulated data. The effect of measurement noise on the network performance was also reported.

Mixed-mode open-notch flexure (MONF), anti-symmetric loaded end-notched flexure (MENF) and centre-notched flexure (MCNF) specimens were used to investigate dynamic mixed I/II mode delamination fracture using a fracturing split Hopkinson pressure bar (F-SHPB) [169]. An expression for dynamic SERR G_d was formulated and evaluated. The experimental results showed that dynamic delamination increased linearly with mode mixing. At low input energy $E_i \leq 4.0$ J, the dynamic (G_d) and total (G_T) SERRs were independent of mixed-mode ratio. At higher impact energy of $4.0 \leq E_i \leq 9.3$ J, G_d decreased slowly with mixed I/II mode ratio while G_T was observed to increase more rapidly. In general, G_d increased more rapidly with increasing delamination than with increasing energy absorbed. For impact energy of 9.3 J before fragmentation of the plate, the effect of kinetic energy was not significant and could be neglected. For the same energy-absorption level, the delamination was greatest at low mixed-mode ratios corresponding to highest Mode II contribution (Figure 29). The results of SERRs from MONF were compared with a mixed-mode bending (MMB) formulation and were in agreement for Mode II but not for Mode I. Hackle (Mode II) features on SEM photographs decreased as the impact energy increased, but increased as the Mode I/II ratio decreased. For the same loading conditions, more pure Mode II features were generated on the MCNF specimen fractured surfaces than the MENF and MONF specimens.

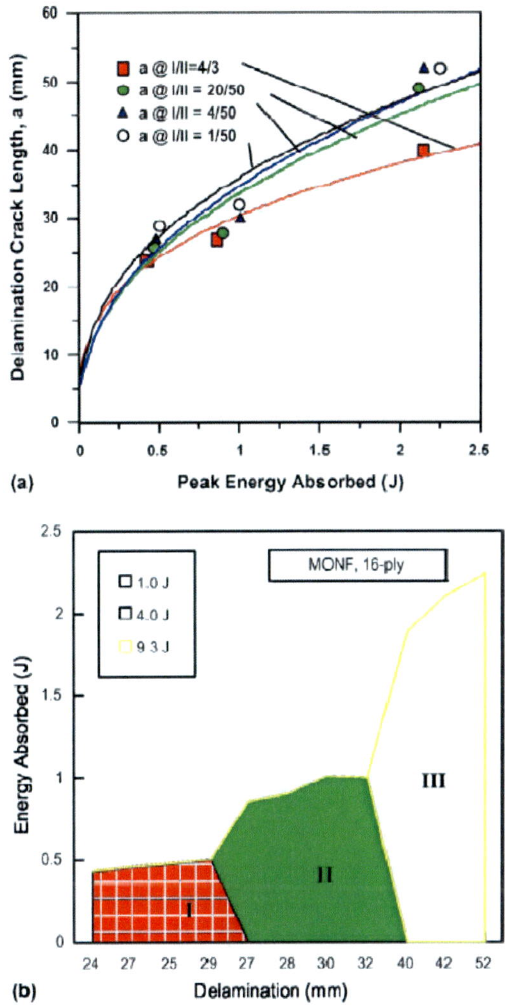

Figure 29. (a) Delamination crack growth as a function of peak energy absorbed at 9.3 J impact Energy for varying mode ratios and (b) regions of crack growth and area plot of energy absorbed-delamination curve for varying impact energy [169].

Another approach to calculate the stiffness reduction due to the delamination presence was proposed by Kim et al [170]. A damage detection problem was formulated as the identification of the spatial stiffness distribution in a damaged composite plate. Full-field heterogeneous curvature fields obtained using an optical deflectometry technique were processed by the virtual fields method adapted to retrieve the 2D stiffness distribution map of a damaged carbon-epoxy plate (Figure 30). The method located the damage and provided a fairly good estimate of the stiffness reduction in the damaged area. In this paper, the procedure was described, validated on simulated measurements and some initial experimental results were given.

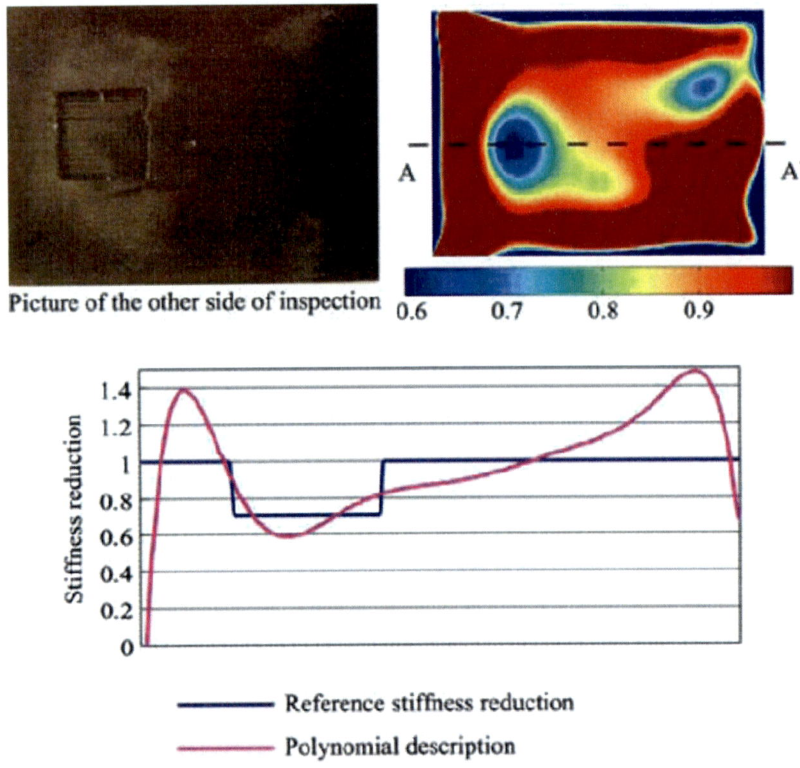

Figure 30. Optical deflectometry curvature fields processed with the virtual field method to retrieve the stiffness distribution [170].

An analytical model for the prediction of compressive fatigue threshold strains in composite plates with barely visible impact damage (BVID) was presented [171]. The model represented the complex damage morphology as a single circular delamination at a critical level and calculated the strain at which thin-film buckling of the circular delaminated region occurred. The threshold strain was defined as the strain at which the SERR for the fracture of post-buckled delaminated plies, along the delamination, was equal to the critical Mode I value (G_{IC}) for the resin. The model predicted the critical through-thickness level for delamination, the stability of delamination growth and the sensitivity to experimental error in geometric measurements of the damage area. Results obtained using the model were in agreement to within 4% for fatigue strain experimental values for four sets of data reported in the literature.

Chapter 7

TECHNOLOGIES FOR THE REDUCTION OR ELIMINATION OF DAMAGE PROPAGATION

In the past, various methodologies have been proposed, studied and used to hinder interlaminar damage and delamination. These include matrix toughening procedures and through thickness laminate reinforcement. The later could be divided into two categories, Z-pinning and stitching.

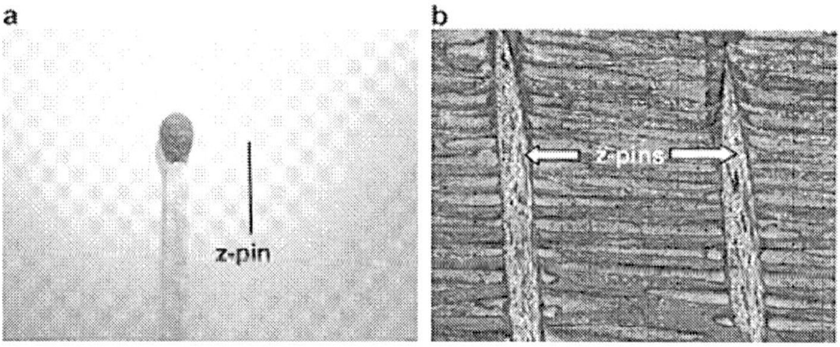

Figure 31. (a) Typical size of a z-pin and (b) z-pins inside a composite [172].

z-Pinning

Mouritz [172] reviewed published research into polymer composite laminates, reinforced in the through-thickness direction with z-pins (Figure 31). The current research into the manufacture, microstructure, delamination resistance, damage tolerance, joint strength and mechanical properties of z-pinned composites was outlined. The benefits of reinforcing composites with z-pins were assessed, including improvements to the delamination toughness, impact damage resistance, post-impact damage tolerance and through-thickness properties. The enhancement of the failure strength of bonded and bearing joints due to z-pinning was also examined (Figure 32). The paper reviewed research into the adverse effects of z-pins on the in-plane mechanical properties, such as reduced elastic modulus, strength and fatigue performance. Mechanisms, responsible for the reduction of the in-plane properties, were discussed and techniques to minimise the adverse effect of z-pins were described. The benefits and drawbacks of z-pinning on interlaminar toughness, damage tolerance and in-plane mechanical properties were compared against other common types of through-thickness reinforcement for composites, such as 3D weaving and stitching. Gaps in our understanding and unresolved research problems with z-pinned composites were identified to provide a road map for future research into these materials.

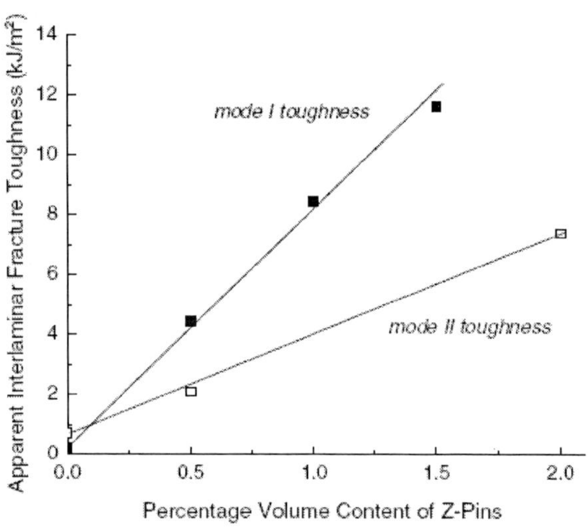

Figure 32. Plots of the effects of z-pins on the apparent modes I and II delamination toughness for CFRPs [172].

Partridge and Cartié [173] provided an introduction into the technology of through-the-thickness reinforcement of thermosetting composites by the insertion of z-fibre pins. The manufacture of the raw materials was described, as was the method of pin insertion. Delamination tests samples were prepared from unidirectional continuous CF/epoxy prepreg (IMS/924), made into 3 mm thick unidirectional laminates with and without a block of z-pins in the crack path. Fracture testing was carried out under Mode I (standard DCB test) and Mode II (3-point-ENF) loading conditions. In the Mode II loading case, the z-pins underwent a significant bending deformation prior to failure by internal shear [174]. The data presented were obtained by using the commonly used data reduction schemes. The applicability of such data treatment was also critically assessed.

The effectiveness of z-pins in co-cured joints was illustrated on the model of a composite double cantilever beam (DCB) subject to a standard fracture toughness test [175]. A comprehensive solution was presented accounting for a broad spectrum of issues that affect the problem. The accurate evaluation of the rotational constraint provided by the intact section of DCB, possible transverse shear deformation in the delaminated section, and effects of uniform and non-uniform temperature were considered. A simple criterion for the effectiveness of z-pins in co-cured joints was introduced and its application was illustrated on numerous examples. As indicated by the analysis, z-pinning was an effective method of enhancing delamination resistance of composite joints. Even a very small volume fraction of z-pins (less than 1.5%) arrested delamination in co-cured composite joints.

The mode I interlaminar fracture in z-pin reinforced composite laminates was modelled [176] using a cohesive volumetric finite element (CVFE) scheme. The test configuration used in this study was a z-pin reinforced double cantilever beam specimen. A bilinear rate-independent but damage-dependent cohesive traction – separation law was adopted to model the fracture of the unreinforced composite, and discrete nonlinear spring elements were employed to represent the effect of the z-pins. The delamination toughness and failure strength of the z-pin reinforced composites were determined by a detailed comparative study of the numerical modelling results with experimental data. To further reduce the computational effort, an equivalent distributed cohesive model was introduced as a substitute for the discrete nonlinear spring representation of the z-pins. The cohesive model was implemented on various test problems with varying failure parameters and for varying spatial z-pin reinforcement configurations and was in good agreement with experimental results.

Chang et al [177] examined the effect of pinning on the flexural properties, fatigue life and failure mechanisms of carbon/epoxy laminates. Five-harness satin weave carbon/epoxy laminates were reinforced in the through-thickness direction with different volume fractions and sizes of fibrous composite pins. Microscopic examination of the laminates before flexural testing revealed that the pins caused considerable damage to the microstructure, including out-of-plane crimping, in-plane distortion and breakage of the fibres and the formation of resin-rich zones around each pin. The pins also caused swelling of the laminate that reduced the fibre volume content. Despite the damage, the pins did not affect the flexural modulus of the laminate as was previously reported by Chang et al [178]. However, increasing the volume content or the diameter of the pins caused a deterioration in the flexural strength and fatigue life, which appeared to be governed by fibre rupture on the tensile side of the laminate. Property changes were discussed in terms of transitions in the dominant failure mechanisms due to the presence of pins.

The effect of z-pinning on the in-plane compression properties and failure mechanisms of polymer laminates was experimentally studied by Mouritz [179]. The reduction of the compression modulus, strength and fatigue performance of carbon/epoxy laminates with increasing volume content and diameter of pins was determined. The elastic modulus decreased at a quasi-linear rate with increasing pin content and pin diameter. Softening was caused by fibre waviness around the pins and reduced fibre volume content due to volumetric swelling of the laminate from the pins. A simple model was presented for calculating the compression modulus of pinned laminates, which took into account the softening effects of fibre waviness and fibre dilution. The compression strength and fatigue life also decreased with increasing volume content and pin diameter. The strength and fatigue properties deteriorated due to fibre kinking caused by fibre waviness around the pins and the reduced fibre content caused by swelling. The deterioration of the compression properties was also dependent on the laminate lay-up. The decrease in properties was dependent on the percentage of 0° or load bearing fibres in the laminate. The paper provided suggestions for minimizing the loss of compression properties to laminates due to pinning.

Stitching

Carpino et al [104] claimed that stitching technologies were particularly effective in improving damage tolerance. Stitching is expected to play an important role in primary structures of next generation commercial airplanes. However, they postulated that stitching could be detrimental to some mechanical properties, such as quasi-static in-plane stiffness and strength. Perforation energy was also adversely affected by stitching. Carpino et al [104] explained the detrimental effects of stitching as an effect of the synergies of stitching in combination with sample-bullet contact and performed the required numerical analyses for their description and modelling. In their review [104], dynamic tests were performed on stitched CFRP panels of different thicknesses. The panels were struck by two spherical projectiles of different diameter. The experimental results were in agreement with two closed-form models for the prediction of the perforation energy, one of which was suitably modified. The damaged area of the impacted samples was recorded by ultrasonic C-scan, in order to study the correlation between delamination extent and the maximum energy absorbed by the target during impact.

Experimentally measured perforation data were given for woven, z-stitched and through-thickness z-stitched glass polyester laminates for a number of laminate thicknesses, a number of geometries of impactor (cone, flat and hemisphere), and two missile masses [180]. Impact perforation velocities ranged up to 571 m s^{-1} on 200×200 mm laminates with fully clamped boundary conditions (Figure 33). Results were expressed in terms of static and impact perforation energies. The discussion included a study of energy absorption mechanisms during perforation, with a view to identifying improved combinations of materials. It was concluded that all configurations behave in a similar manner and that the flat ended missile had the largest dynamic enhancement factor, i.e. ratio of impact perforation energy to static perforation energy.

Aymerich et al [181] studied the effect of edge stitching on tensile static and fatigue properties of graphite fibre reinforced laminates. In a following work [182], they examined the effect of stitching on the impact performance of a class of graphite/epoxy cross-ply laminates with the aim of investigating the ability of through-thickness reinforcement to improve delamination resistance. The influence of Kevlar® stitching on edge delaminations, ultimate tensile strength and tension–tension fatigue life was examined on laminates of two stacking sequences ([±45/0/90]s and [±30/90]s), as representative of two classes of fibre-dominated and matrix-dominated laminates. Through-thickness stitching, offered a significant improvement in the static delamination resistance of laminated composites, but had adverse effects on the ultimate tensile stress of the laminates under study, with an increase in strength of $[\pm 30/90]_s$ laminates and a decrease in strength of $[\pm 45/0/90]_s$ laminates. Similarly the fatigue life of matrix dominated $[\pm 30/90]_s$ specimens was considerably extended by stitching, while the fatigue resistance of fibre-dominated $[\pm 45/0/90]_s$ specimens was reduced, particularly at high fatigue stresses. Stitching appeared very efficient in arresting delamination in

$[\pm 30/90]_s$ specimens for both static and fatigue loads, but was not as efficient in stopping fatigue delaminations in $[\pm 45/0/90]_s$ specimens, since Kevlar® threads were found to be prone to breakage under cyclic loading. While the static delamination resistance of laminated composites was improved by stitching, this did not automatically result in better fatigue performance for all lamination sequences.

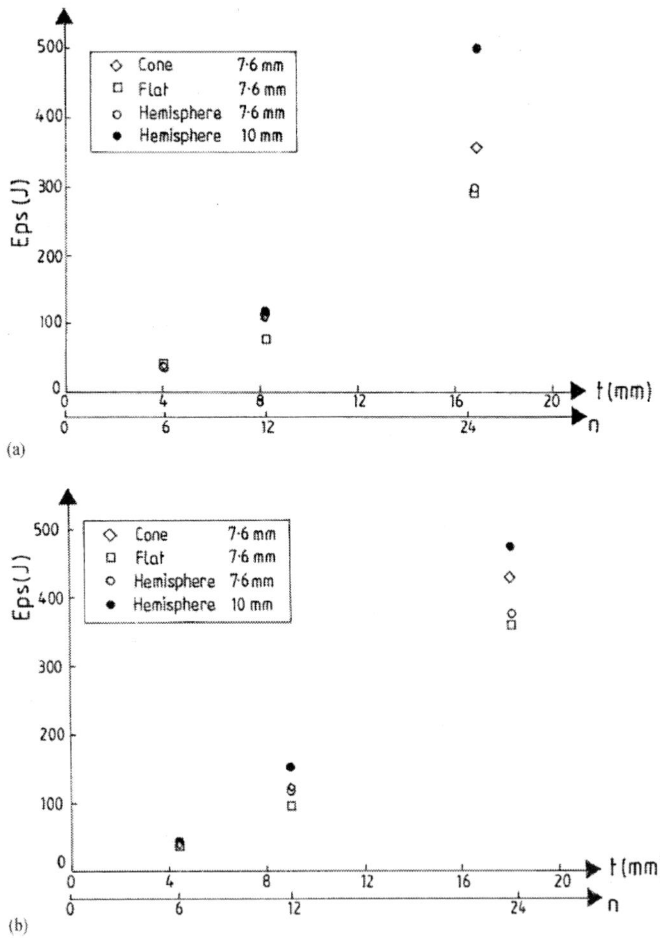

Figure 33. Static energy to perforation versus thickness and number of plies for (a) Woven roving laminates and (b) z-stitched laminates [180].

In the work by Hosur et al [183], damage resistance of stitched/unstitched S2-glass/epoxy composites was studied. Five layer stitched/unstitched plain weave S2-glass woven fabric composite laminates were manufactured using a toughened SC15 epoxy resin system. For stitching, two configurations: one with 25.4 mm grid and other with 12.7 mm grid, were used with 6 mm pitch. Damage resistance was evaluated by subjecting 100×100 mm^2 samples to LVI loading at energy levels ranging from 10 J to 80 J using an instrumented impact tester. Three samples were tested at each energy level. The extent of damage was evaluated using ultrasonic C-scan. The outcome of the study indicated that stitching confined the damage size. As damage was little over most of the energy range, the study was further extended to determine the effect of repeated impact loading. Under this study, laminates were subjected to repeat impact loading up to a maximum of 40 impacts at energy levels ranging from 10 to 50 J. Results of the repeated impact study were reported in terms of peak load, absorbed energy and projected damage area. All the laminates, sustained repeated impact loading up to 30 J. Beyond 30 J, laminates failed to carry repeated loading and failure depended on the laminate configuration.

Saito and Kimpara [184] focused on a multi-axial stitched fabric, which was a thick, high performance reinforcement for large-scale composite structures. The effects of impact damage on multi-axial stitched CFRP laminates moulded by vacuum-assisted resin transfer moulding (VARTM) method were evaluated. Impact damage within the material was evaluated by an ultrasonic scanning device and optical cross-sectional observations. Probed images, obtained by both non-destructive and destructive methods, were compared and internal damage distributions of multi-axial stitched CFRP laminates were compiled. In addition, the residual compressive strength and the fatigue properties of impact-damaged CFRP laminates were evaluated by in situ damage growth monitoring using the thermo-elastic stress analyzer (TESA). Three-dimensional damage distributions of impacted CFRP laminate were obtained from ultrasonic C-scan images and cross-sectional photographs. The damage progress behaviour was observed on a destructive and non-destructive basis by post-impact-fatigue (PIF) tests.

LVI tests were carried out [185] on stitched CF-reinforced plastic laminates of various thicknesses, with reference to the overall force – displacement curve, first failure load, penetration, indentation and damage extent. The results obtained were compared to similar data available for 2D laminates. Apparently, the presence of stitches did not affect substantially the material behaviour in terms of force – displacement curve, first failure load, and indentation. However, the stitched laminates exhibited penetration energy about 30% lower than their 2D equivalent. The advantage of stitching in terms of impact damage resistance was evident only for high thickness composites. The data generated suggested that the use of stitches could be unnecessary to hinder delamination in thin 2D laminates.

Velmurugan and Solaimurugan [186] examined the effect of plain stitching by untwisted fibre rovings in the in-plane mechanical properties and Mode I interlaminar fracture toughness of glass/polyester composites. Through-the-thickness reinforcement was performed by stitching with twisted fibre yarns in a sewing machine to increase the delamination strength. However, stitching degraded the in-plane mechanical properties. In this work glass woven roving mats (WRM) were stitched in the thickness direction with untwisted Kevlar®, glass and carbon fibre rovings by plain stitch. It was observed that in plain stitch, no thread cross was formed and thus no resin rich pockets were seen. The uniform distribution of fibres in the stitch roving, the absence of resin rich region and reduced fibre damage resulted in increased in-plane tensile, lap shear, flexural, transverse shear and impact strengths. The effect of stitching on Mode I delamination toughness (G_{Ic}) of glass/polyester laminates was investigated by performing double cantilever beam (DCB) test. Stitching increased the Mode I delamination toughness up to 20 times compared to that of unstitched specimens.

The influence of pure assembly seams, using a thin polyester yarn in a zigzag geometry, on the resulting mechanical performance of a non-crimped fabric (NCF) CF-reinforced epoxy composite, manufactured by vacuum-assisted resin transfer moulding, was presented by Beier et al [187]. This study aimed at generating a solid foundation regarding the overall performance level of stitched NCF composites and at identifying critical property changes. The comprehensive evaluation of the mechanical composite properties included static as well as dynamic tests of the in-plane properties as well as a characterisation of the interlaminar properties, such as apparent interlaminar shear strength (ILSS) and CAI. It was demonstrated that mechanical properties such as the tensile and compression stiffness and CAI strength were not degraded by the chosen stitching parameters, whereas the tensile and compression strength, ILSS as well as the tensile fatigue behaviour were reduced as a result of pronounced localised fibre undulations. A direct comparison with the properties of a commonly used 5H satin woven fabric composite verified that the overall performance of the stitched NCF composites were superior with regard to the key criteria for aircraft applications and maintained the performance advantage of NCF composites as compared to standard woven fabrics. Promising approaches included the use of different yarn materials based on soluble thermoplastics and/or modified stitching parameters.

The tensile fatigue properties of specific types of 3D woven, stitched and z-pinned composites with through-thickness reinforcement, were compared by Mouritz [188]. Tensile tests under monotonic and cyclic loading were performed on the 3D composite materials to determine the influence of the z-reinforcement type – woven z-binder, stitch or z-pin – on the tensile modulus, strength and fatigue life. The in-plane Young's modulus of the composites was not affected by the type or volume content of the z-reinforcement. The tensile strength of the 3D woven and stitched composites was also not affected by the z-reinforcement. However the strength of the z-pinned composite dropped steadily with increasing volume content of z-reinforcement. The fatigue life of the 3D composites was reduced by the z-reinforcement, regardless of the type of z-reinforcement, i.e. woven z-binders, stitches or z-pins. Z-reinforcement damaged the microstructure of the 3D composites and resulted in the degradation of the tensile fatigue properties. The fatigue damage mechanisms caused by the different types of z-reinforcement were described. The results indicated that through-the-thickness reinforcement was detrimental to the tensile fatigue life. However, the study was restricted to specific types of materials and further research into a wider variety of 3D woven, stitched and z-pinned composites was required for a general assessment of their fatigue performance.

Tan et al [189] investigated the effect of selective stitching on impact damage tolerance and fatigue durability of stiffened composite panels and compared it with unstitched stiffened composite panels. Two-blade stiffened composite panels were fabricated by the resin film infiltration technique. Impact damage was inflicted on the stiffened panels using a drop-weight with impact energy of 30 J from the skin-side over the stiffener and the flange. The experimental results indicated that selective stitching was an effective way to improve the compressive failure strength and fatigue strength of stiffened panels with both clearly visible flange damage (CVFD) and clearly visible stiffener damage (CVSD). The buckling and post-buckling FEA were performed to predict the static compressive strength of the selectively stitched stiffened panels. A good agreement was obtained between analytical and experimental results. The failure behaviour of the undamaged, CVFD and CVSD panels was also investigated.

Matrix Toughening – Interleaving

Matrix toughening procedures can be categorized into traditional interleaving reinforcement and the incorporation of nano-phases to produce hybrid composites.

Kishore et al [190] examined the dynamic behaviour of laminated carbon/epoxy composites with inserted polytetrafluoroethylene (PTFE)-coated interleaves. Instrumented impact tests performed on the interleaved test samples showed significant differences in the energy absorption characteristics that could be correlated with the failure mode. It was inferred that with the introduction of tougher layers of material at specific locations, the load-carrying ability decreased while the energy-absorbing capability improved considerably.

The microstructure and fracture behaviour of epoxy mixtures containing two monomers of different molecular weights were studied by Zubeldia et al [191]. The variation of fracture toughness by the addition of other modifiers was also investigated. Several amounts of high-molecular-weight diglycidyl ether of bisphenol A (DGEBA) oligomer were added to a nearly pure DGEBA monomer. The mixtures were cured with an aromatic amine and exhibited phase separation after curing. The curing behaviour of the epoxy mixtures was investigated with thermal measurements. A significant enhancement of the fracture toughness was accompanied by a slight increase in both the rigidity and strength of the mixtures that corresponded to the content of the high-molecular-weight epoxy resin. Dynamic mechanical analysis and atomic force microscopy studies indicated that the generated two-phase morphology was a function of the content of the added epoxy resin. The influence of the addition of an oligomer or a thermoplastic on the morphologies and mechanical properties of both epoxy-containing mixtures was also investigated.

Hybrid composite materials with non-woven tissue (NWT) were developed to improve the mechanical properties of conventional FRP composite materials [192]. As was shown, the interleaving of the non-woven carbon tissue (NWCT) layer largely increased the Mode II interlaminar fracture toughness while it did not significantly change the Mode I interlaminar fracture toughness. In addition, compared with conventional CFRP composites, the hybrid composites with NWCT exhibited high reliability and improved strength in their response to quasi static tension and fatigue. The low-energy impact damage mechanism for the NWCT hybrid composites was also investigated under quasi-static indentation loading. The toughening mechanism of the NWCT interleaving was described in relation to observed damage. Indentation caused shear failure due to the compressive fatigue loading of the CFRP and hybrid composites, but in the case of hybrid composites, the propagation of shear cracks was prevented by NWCT located between the CFRP layers. It was found that the Mode II interlaminar fracture was predominant in the delamination by the indentation loads.

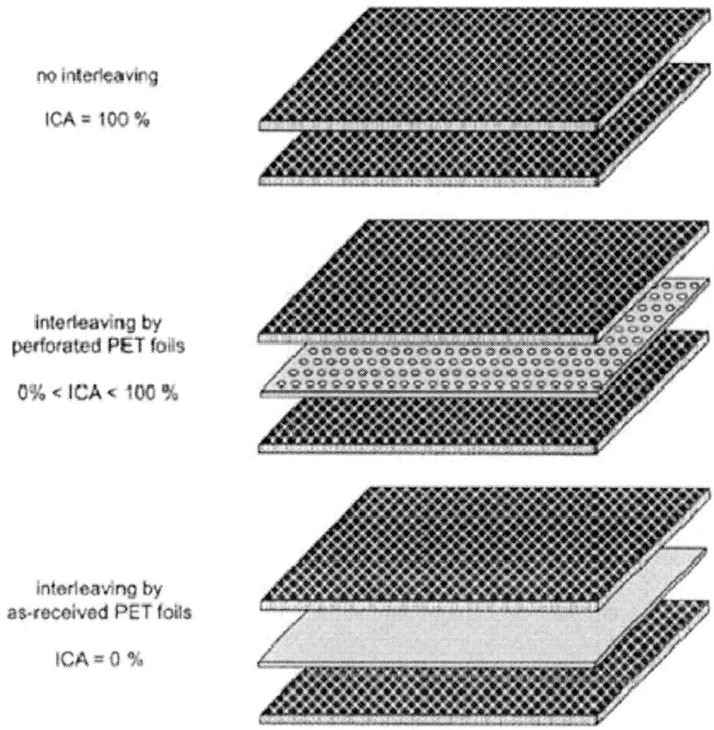

Figure 34. Static sequence for non-interleaved and PET interleaved laminates [193].

Pegoretti et al [193] studied the correlation between interlaminar fracture toughness and impact energy absorption for the fracture of carbon/epoxy laminates. CF/epoxy cross-ply prepreg layers were interleaved with thin poly(ethylene-terephthalate) (PET) films. Before the lamination process, 1 mm diameter circular holes were drilled in the PET films at several surface densities (from 0 up to 44 holes/cm^2) in order to selectively increase the interlaminar contact area between the carbon/epoxy laminae (Figure 34). In this way, the interlaminar contact area gradually varied from 0%, corresponding to the case in which non-perforated PET films were used, up to 100% in the case of non-interleaved laminates. The Mode I interlaminar fracture toughness of the resulting laminates was determined according to the ASTM D-5528-01 standard test method on double cantilever beam (DCB) specimens. The critical values of the SERR determined at the point at which the load versus opening displacement curve became non-linear ($G_{IC,NL}$) varied from 40 up to 260 J/m^2, depending on the interlaminar contact area. All the laminates were subsequently tested in three point flexure performed both under quasi-static (5 mm/min) and impact (2 m/s) loading conditions. The elastic modulus of the laminates was found to be practically independent of the level of interlaminar adhesion, while the bending strength decreased as the interlaminar fracture toughness decreased. The total energy to fracture evaluated under impact conditions showed a non-monotonic correlation with interlaminar fracture toughness, reaching a maximum level for a $G_{IC,NL}$ value of about 60 J/m^2. At the same time, the ductility index, i.e. the ratio between the propagation and the initiation energies, evaluated by instrumented Charpy impact tests, markedly increased as the interlaminar fracture toughness decreased (Figure 35).

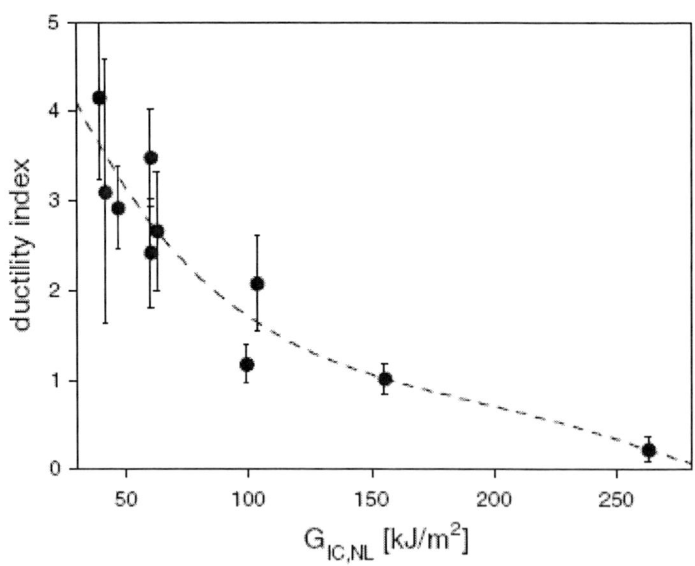

Figure 35. Ductility index values as a function of the interlaminar fracture toughness [193].

Hybrid Composite Systems Incorporating Nano- Phases

Carbon nanotubes (CNTs) are considered to be highly promising as fillers for improving the properties of polymers. However, questions concerning the appropriate type of CNTs, e.g., single-wall CNTs (SWCNT), double-wall CNTs (DWCNT) or multi-wall CNTs (MWCNT), and the relevance of a surface functionalisation are still to be answered. In their study [194] Gojny et al focused on the evaluation of the influence of different types of nano-fillers on the mechanical properties of epoxy-based nano-composites and the relevance of surface functionalisation. The produced nano-composites exhibited enhanced strength and stiffness and more importantly, a significant increase in fracture toughness (43% at 0.5 wt% amino-functionalised DWCNT). The influence of filler content, the dispersion, the aspect ratio, the specific surface area and amino-functionalisation on the nano-composite properties were discussed and correlated to the identified micro-mechanical mechanisms (Figure 36).

Sadeghian et al [195] presented an experimental study of using carbon nanofibres (CNF) to improve the interlaminar fracture properties of GF/polyester composites. Surfactant-treated CNF were dispersed in polyester resin and then the CNF-resin suspension was used to impregnate a GF preform using VARTM. The possibility to use VARTM for thick and large CNF toughened composite parts was experimentally investigated. The influence of CNF concentration on the CNF filtration in the GF preform, the resin viscosity, and the micro-void formation were examined (Figure 37). By choosing appropriate manufacturing parameters, they were able to use the VARTM process to infuse the surfactant-treated CNF/resin matrix into the GF preform and successfully manufacture the CNF toughened GF/polyester specimens for mode-I delamination tests. The critical SERR of mode-I delamination (G_{IC}) was estimated for several composite specimens with 1 wt% CNF concentrations. Significant improvement in the G_{IC} was consistently observed when 1 wt% CNF were added to toughen the polyester resin. Microscopic examination showed that the fracture surface of the 1 wt% CNF toughened GF/polyester samples was more complex than the fracture surface of regular GF/polyester composites.

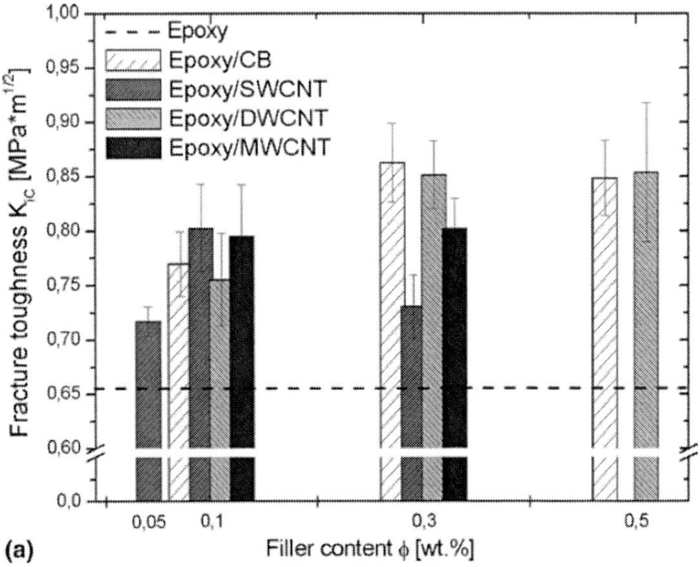

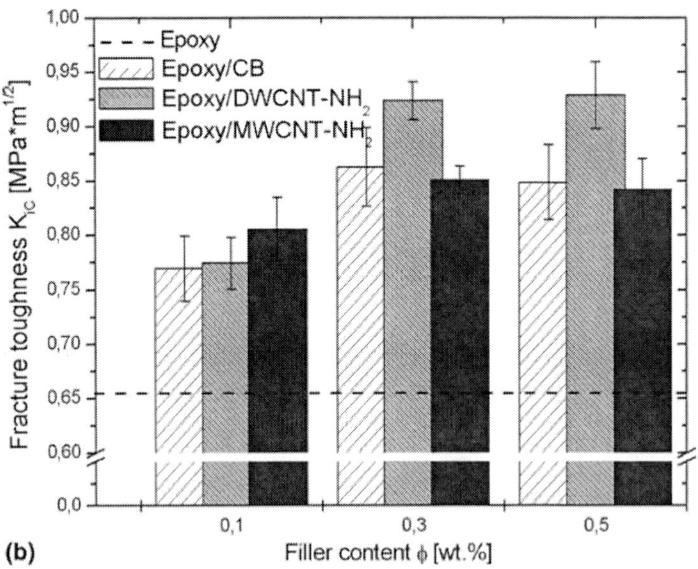

Figure 36. Experimentally obtained fracture toughness of epoxy-based composites containing (a) non-functionalised and (b) amino-functionalised nanoparticles. [194].

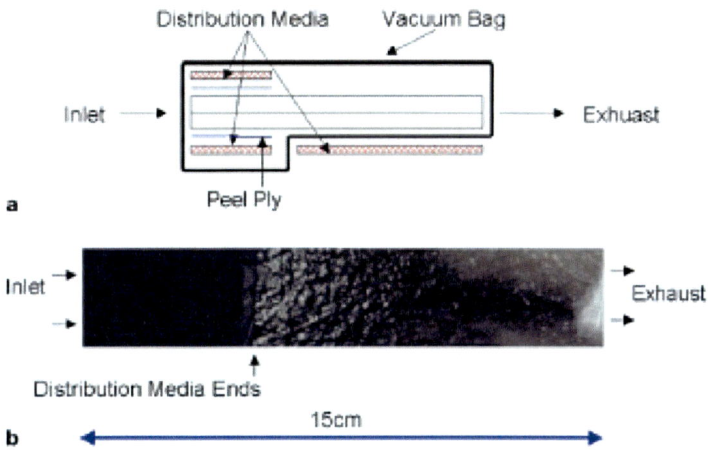

Figure 37. Filtration in flow direction with inflow with 1.0 % CNF concentration (a) perform and setup with two layers of random glass fibre mats with thickness 1.2 mm (side view) and (b) contrast difference due to CNF filtration [195].

Wetzel et al [196] focused on the effect of nano-particles on the mechanical and fracture mechanical properties of epoxy resins, particularly with regard to fracture and toughening mechanisms. A comprehensive study was carried out on a series of nano-composites containing varying amounts of nano-particles, either titanium dioxide (TiO_2) or aluminium oxide (Al_2O_3). Nano-composites were systematically produced by applying high shear energy during a controlled dispersion process, in order to reduce the size of agglomerates and produce a homogeneous distribution of individual nano-particles within the epoxy resin. The mechanical performance of the nano-composites was subsequently characterized in flexure, dynamic mechanical analysis (DMA) and fatigue crack growth testing (FCP), using a linear fracture mechanics approaches (LEFM). The microstructure of specimens and the corresponding fracture surfaces were examined by TEM, SEM and AFM techniques in order to identify the relevant fracture mechanisms, and to gain information about the dispersion quality of nano-particles within the polymer. It was found that the presence of nano-particles in epoxy triggered various fracture mechanisms, e.g. crack deflection, plastic deformation, and crack pinning. It was observed that nano-particles could overcome the drawbacks of traditional tougheners (e.g. glass beads or rubber particles) by simultaneously improving stiffness, strength and toughness of epoxy, without sacrificing thermo-mechanical properties (Figure 38).

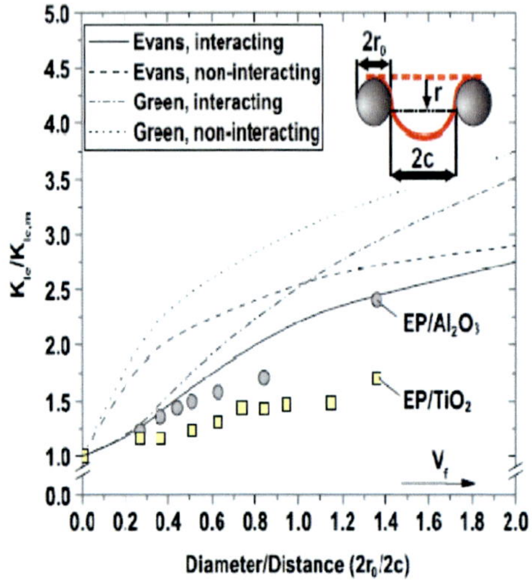

Figure 38. Increase of the fracture toughness with increasing diameter to distance ratio ($2r_0/2_c$) of the fillers. Experimental data compared with theoretical models for crack pinning [196].

Mode I and II interlaminar fracture toughness were investigated [197] for CF/epoxy laminates toughened by a carbon nanofibre/epoxy interlayer. Vapour grown carbon fibre (VGCF) and vapour grown carbon nanofibre (VGNF) were chosen as the reinforcement for the interlayer. To illustrate the effect of the interlayer on the fracture toughness of the laminates, several types of CF reinforced plastics/carbon nanofibre (CFRP/CNF) hybrid laminates were fabricated. The laminates were manufactured using unidirectional carbon/epoxy prepregs and a carbon nanofibre interlayer with varying thickness. The mode I interlaminar fracture toughness was evaluated by a standard double cantilever beam (DCB) test. The mode II interlaminar fracture toughness was evaluated by the end notched flexure (ENF) test. The DCB tests confirmed that the mode I interlaminar fracture toughness for hybrid laminates was about 50% greater than that of the plain CFRP laminates. Furthermore, the mode II fracture toughness tests confirmed that the interlaminar fracture toughness for hybrid laminates was 2–3 times greater than that of plain CFRP laminates. The recommended range of CNF interlayer thickness was between 100 and 150 μm (approximately 20 g/m^2 carbon nanofibre area density) (Figure 39). The difference in the effect of the interlayer fracture properties under mode I and II fracture was discussed on the basis of SEM fractographic observations.

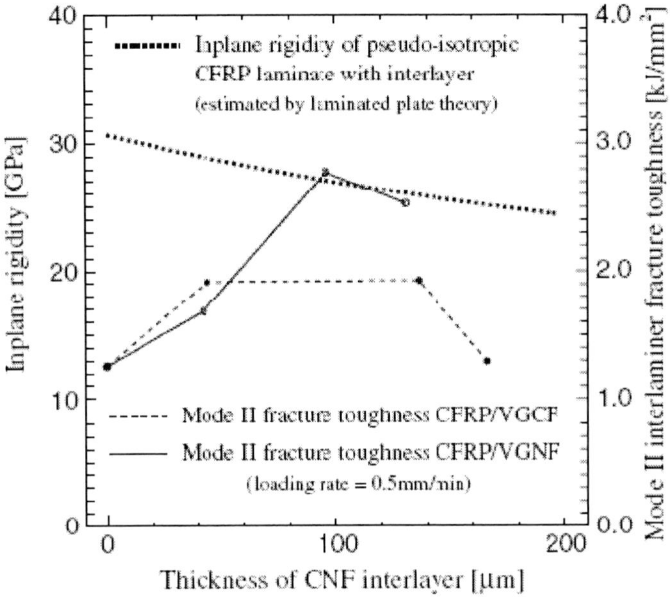

Figure 39. In plane rigidity of quasi isotropic CFRP laminate and Mode II interlaminar fracture toughness versus CNF interlayer thickness [197].

Natural Composites

The influence of the fibre-matrix adhesion on the fatigue and impact performance of jute fibre-reinforced polypropylene was investigated [198]. It was shown that a strong interface was connected with a higher dynamic modulus and stiffness degradation with increasing load cycles and maximum applied stress. The specific damping capacity was higher for composites with poorly bonded fibres. Furthermore, stronger fibre-matrix adhesion reduced the loss energy for non-penetration impact tests by approximately 30%. Tests performed at different temperatures, exhibited higher loss energies for cold and warm test conditions compared to room temperature tests. The post-impact dynamic modulus (after 5 impact events) was approximately 40% and 30% lower for composites with poor and good fibre-matrix adhesion, respectively.

Chapter 8

APPLICATIONS

A new processing method was developed for spreading fibre tows to make thin-ply laminated composites. The proposed method employed a constant airflow through sagged fibre filaments to efficiently spread the thick tows without damaging any fibres. The method was robust and provided comparable outcome with other available thin-ply methods. The thin plies of thickness less than one-third of the conventional plies could easily be made with the tow-spreading technology. Experiments were performed [199] to evaluate the performance of tow-spread, thin-ply laminated composites. To study the thickness effect of the laminated composites, the test specimens were made with the same material and the same spread tows, but with dispersed and grouped laminations of the plies. Uniaxial tension static and fatigue loadings were applied on both unnotched and open-hole specimens. Impact and compression-after-impact tests were also conducted. From stress–strain curves, acoustic emission counts, X-ray photos, c-scan images and observation of damage modes of failed specimens, it was observed that the thin-ply laminated composites can suppress the micro-cracking, delamination and splitting damage for static, fatigue and impact loadings without special resin and/or 3-D reinforcements. Therefore, the laminate design could be simplified by using higher strain allowables without a progressive failure analysis.

COMPOSITE JOINTS

Lanciotti et al [200] described the results of a research activity, carried out at the Department of Aerospace Engineering at the University of Pisa in conjunction with AGUSTA, aimed at the characterisation of the fatigue behaviour of composite joints. The activity was performed as a support to the certification of the tail of the new helicopter EH 101. Two types of specimens were defined as representative of the structure; the specimens differed in the thickness of the laminates, namely 2 and 4 mm, and were representative of critical areas. The specimens were impact damaged with an energy level capable of inducing barely visible indentation, as required by the Certification Authority. Fatigue tests, under sinusoidal loading, R=O-2, were carried out in both dry and wet conditions; the stiffness of the specimens was monitored during the fatigue tests. The comparison between the results obtained showed the considerable effect of impact damage and of humidity absorption on the fatigue resistance of composite joints.

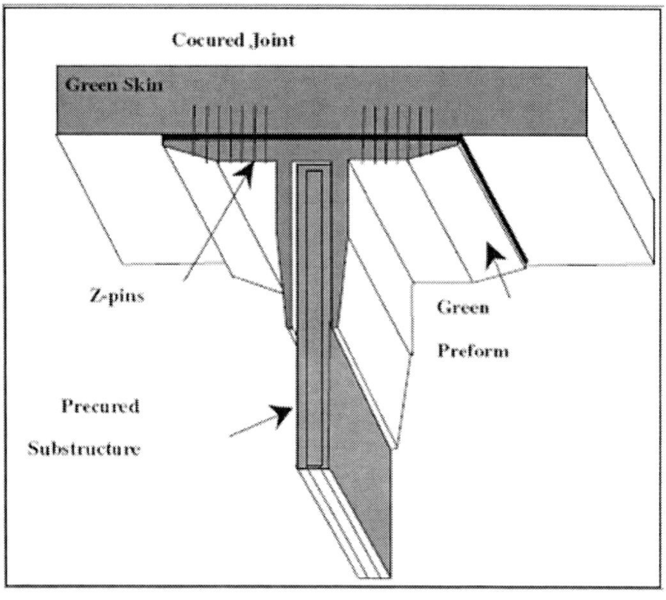

Figure 40. C-cured z-pinned joint between the skin and the stiffener [201].

Although z-pins were effective in preventing delaminations in adhesively bonded and co-cured joints, their applicability depends on the reliable assessment of the strength of a z-pin-composite assembly (Figure 40). In particular, high residual thermal stresses experimentally observed dictated the necessity for local stress analysis. Elevated temperature experienced by the joint during its lifetime may also affect its resistance in delamination. Byrd and Birman [201] illustrated an approach for determining local residual stresses and confirmed the possibility of delamination and cracking in the composite structure due to high post-processing transverse stresses. The analysis of the effect of elevated temperature applied at one of the surfaces on the response of a z-pinned joint was conducted using the concept of a double cantilever beam with an "insulated" crack. In addition, it was shown that elevated temperature could actually enhance the integrity of the joint provided that the z-pin-composite interfacial strength increased.

Wind Turbines

The size of wind turbine rotors has increased in the past decade from 40 m to more than 120 m diameter. The resulting mass of about 18 tons per rotor blade causes high bending moments at the inner part of a blade due to the gravitational loads. More than 10^8 load cycles will take place in the expected lifetime of 20 years of a turbine. During this time the rotor blades are exposed to various hostile conditions such as extreme temperatures, humidity, rain, hail impact, snow, ice, solar radiation, lightning and salinity. In order to withstand these external conditions without diminishing the safety, a sound knowledge of the fatigue behaviour of the material and structural properties is needed. To meet the upcoming requirements Kensche [202] highlighted some fatigue and lifetime aspects on wind turbine rotor blades made of composite materials. These included a historical part in connection with glider technology, the presentation of relevant $S-N$ curves not only for the 0°-orientated fibres representing the spar cap but also for ±45°-lay-ups in shear web and shell, the influence of fibre content and architecture, of environmental effects, with a view to lifetime prediction of structural elements.

Fibre Metal Laminates

Fibre metal laminates offer significant improvements over current available materials for aircraft structures. While weight reduction and improved damage tolerance characteristics were the prime drivers to develop this new family of materials, they also possess additional advantages, critical for today's designers, such as cost reduction and improved safety. Originally developed for their outstanding fatigue resistance, other advantages of fibre metal laminates include high specific static properties, ease of manufacture, excellent impact resistance, burn through capabilities rivalling titanium alloys, and good corrosion resistance. The combination of these properties in one material was an extraordinary achievement, rendering Glare [203] particularly inviting for aircraft applications. The very nature of Glare is its crack bridging mechanism, which provides superior damage tolerance properties. Depending on the property, Glare shows either monolithic metal or composite behaviour, which challenges the definition of strength justification and certification procedures. Through its unique combination of properties, Glare was chosen as the material for pressurized fuselage skin structures of the new aircraft generation, such as the A380.

Glare exhibited excellent resistance to crack growth [203]. This was attributed to the combination of delamination and fibre bridging mechanisms. The fatigue insensitive fibres restrained the crack opening and transferred load over the crack in the metal layers. During the initiation phase fibre bridging did not occur and the behaviour was dominated by the metal properties. Mechanically fastened joints introduced additional effects such as secondary bending, load transfer and effects related to the fastener installation. The residual strength of Glare depended on the amount of broken fibres and the delamination size and was described with the R-curve approach.

The impact resistance of Glare was related to the aluminium and glass/epoxy properties and was significantly higher than the impact resistance of monolithic aluminium [204]. The same was proven for fire resistance. Depending on the Glare grade and thickness, the outer aluminium layer melted away, whereas the other layers remained intact due to carbonisation of the glass/epoxy layers and delamination of the laminate. The air in the delaminations acted as insulation, keeping the temperatures at the non-exposed side relatively low.

Some aspects of the detailed design of aircraft structures in Glare, as well as the design of splices and riveted joints were discussed [205]. In order to apply Glare in very large fuselage panels, a splice concept was developed, which allowed for a number of longitudinal splices to be cured in the same curing cycle as the basic material. Through the introduction of this splicing concept, the width of a panel was no longer limited to the maximum width of the aluminium sheet. Internal local reinforcements (doublers) may be integrated into the panel during lay-up.

SUMMARY

An overview of the post-impact-fatigue behaviour of composites was performed. The response of impacted structural composites to service loading is critical as impact events may significantly degrade the material properties.

This work initially focuses on the impact response of composite materials and their constituent phases as they are documented in the literature. The effect of polymer matrix materials, laminate geometry and reinforcement geometry are reported, including sandwich structures, and a thorough account of the experimental, and semi-empirical approaches is provided for the quantification of damage. The effect of temperature and thermal stresses is also assessed. A thorough investigation of the effect of impact-fatigue on the composite properties is performed together with the experimental and modelling approaches that appear in the literature.

Delamination growth under fatigue loading for impacted composite panels is also examined. Delamination growth is assessed in the literature both under mode I and II conditions, both numerically and experimentally. Finally, the models for fatigue life prediction are presented, accounting for the presence of notches, matrix toughness, loading modes and stiffness reduction.

The non destructive evaluation of initial damage and its propagation is of primary importance for composite structures. To this end, the up-to-date methods are presented. These methods are generally based on wave propagation, electric conductivity or via the incorporation of sensors in the parent material, such as Fibre Bragg gratings. Full field techniques, such as optical and lock-in thermography, speckle pattern interferometry and x-radiography are also widely used in the literature.

The post impact response both under static and cyclic loading is also examined. The extended literature on the subject includes among others the effect of notches, the effect of hydrothermal exposure, hail damage as well as the effect of different geometries and fabrication processes with a view service life prediction and/ or optimisation of the composite.

The post-impact behaviour of composites is extensively modelled in the literature. The associated models include semi empirical analyses, numerical studies in conjunction with experimental findings and analytical solutions. Special attention was given to buckling and post-buckling analyses of plates in the presence of delamination. The existing models focus on the different fracture modes as well as the prediction of the critical strain energy release rate.

A thorough investigation of the technologies that inhibit damage initiation and propagation is also presented. These technologies include z-pinning and stitching, interleaving, and matrix toughening. Special focus is given to novel technologies where the incorporation of phases at the nanoscale is employed to increase the toughness of the composite and subsequently its damage tolerance.

Throughout this work, detailed reference to the impact and post-impact behaviour of natural fibre composite is made.

Finally, an overview of the structural applications where increased damage tolerance is a prerequisite is presented with a special mention to fibre metal laminates.

REFERENCES

[1] Sharma, A. V. In *Test Methods and Design Allowable for Fibrous Composites;* Chamis, C.C.; Ed.; ASTM STP 734; American Society for Testing and Materials: Philadelphia, PA, 1981; pp 54–70

[2] Garg, A. C. *Eng Fract Mech.* 1988, *29*, 557–584.

[3] Abrate, S. *Appl Mech Rev.* 1991, *44*, 155–190.

[4] Prasad, C. B.; Ambur, D. R.; Starnes, J. H., Jr. *AIAA Jrnl.* 1994, *32*, 1270–1277.

[5] Markopoulos, Y. P.; Kostopoulos, V. *Adv Comp Lett.* 2003, *12*, 177–190.

[6] Katerelos, D.T.G.; Varna, J. (2008). Secondary Damage Modes in Laminates Initiated by Intralaminar Cracks. Composite for Sustainable Progress. Proceedings of the 13th European Conference on Composite Materials, Stockholm, Sweden, June 2-5, 2006. Electronic Proceedings.

[7] Perkins, W.G. *Polym Eng Sci.* 1999, *39*, 2445-2460.

[8] Arogn, A.S.; Cohen, R.E. *Polymer* 2003, *44*, 6013-6032.

[9] Chotard, T.J.; Benzeggagh, M.L. *Comp Sci Tech.* 1998, 58, 839-854.

[10] Park, R.; Jang, J. *Jrnl Appl Pol Sci.* 2000, *75*, 952-959.

[11] Dutta, P.K.; Hui, D.; Kadiyala, S.V. *Comput Struct.* 2000, *76*, 173-181.

[12] Kim, J.-K.; Sham, M.-L. *Comp Sci Tech.* 2000, 6*0*, 745-761.

[13] Williams , K.V.; Vaziri, R. *Comput Struct.* 2001, *79*, 997-1011.

[14] Morozov, E.V.; Sylantiev, S.A.; Evseev, E.G. *Comp Struct.* 2003, *62*, 367-371.

[15] Corum, J.M.; Battiste, R.L.; Ruggles-Wrenn, M.B. *Comp Sci Tech.* 2003, *63*, 755-769.

[16] Katerelos, D.G.; Paipetis, A.; Kostopoulos, V. *Fat Fract Eng Mater Struct.* 2004, *27*, 911-922.

[17] Cantwell, W.J. *Comp Sci Tech.* 2007, *67*, 1900-1908.
[18] Kang, K.-W.; Kim, H.S.; Kim, M.-S.; Kim, J.-K. *Mater Sci Eng A* 2008, *483-484*, 333-335.
[19] Rilo, N.F.; Silva Ferreira, L.M. *Int Jrnl Mech Mater Des.* 2008, *4*, 291-300.
[20] Im, K.-H.; Cha, C.-S.; Kim, S.-K.; Yang, I.-Y. *Comp B* 2001, *32*, 669-682.
[21] Gómez-del Río, T.; Zaera, R.; Barbero, E.; Navarro, C. *Comp B* 2005, *36*, 41-50.
[22] Sabeel Ahmed, K.; Vijayarangan, S.; Kumar, A. *Jrnl Reinf Plast Comp.* 2007, *26*, 959-976.
[23] Kawaguchi, T.; Nishimura, H.; Ito, K.; Sorimachi, H.; Kuriyama, T.; Narisawa, I. *Comp Sci Tech.* 2004, *64*, 1057-1067.
[24] Chang, L. *Jrnl Mater Sci Lett.* 1985, *4*, 1221-1224.
[25] Chang, L. *Jrnl Mater Sci Lett.* 1985, *4*, 1429-1433.
[26] Roy, R.; Sarkar, B.K.; Bose, N.R. *Comp A* 2001, *32*, 871-876.
[27] Azouaoui, K.; Rechak, S.; Azari, Z.; Benmedakhene, S.; Laksimi, A.; Pluvinage, G. *Int Jrnl Fat.* 2001, *23*, 877-885.
[28] Barkoula, N.-M.; Papanicolaou, G.C.; Karger-Kocsis, J. *Comp Sci Tech.* 2002, *62*, 121-130.
[29] Schrauwen, B.; PeijsT. *Appl Comp Mater.* 2002, *9*, 331-352.
[30] Williams, K.V.; Vaziri, R.; Poursartip, A. *Int Jrnl Sol Struct.* 2003, *40*, 2267-2300.
[31] Sugun, B.S.; Rao, R.M.V.G.K. *Jrnl Reinf Plast Comp.* 2004, *23*, 1583-1599.
[32] Meo, M.; Vignjevic, R.; Marengo, G. *Int Jrnl Mech Sci.* 2005, *47*, 1301-1325.
[33] de Morais, W.A.; Monteiro, S.N.; d'Almeida, J.R.M. *Comp Struct.* 2005, *67*, 307-315.
[34] de Morais, W.A.; Monteiro, S.N.; d'Almeida, J.R.M. *Comp Struct.* 2005, *70*, 223-228.
[35] Rahul Kumar, C.; Radhakrishna, K.; Rao, R.M.V.G.K. *Jrnl Reinf Plast Comp.* 2005, *24*, 949-960.
[36] Baucom, J.N.; Zikry, M.A. *Comp A* 2005, *36*, 658-664.
[37] Baucom, J.N.; Zikry, M.A. *Jrnl Comp Mater.* 2004, *37*, 1651-1674.
[38] Baucom, J.N.; Zikry, M.A.; Rajendran, A.M. *Comp Sci Tech.* 2006, *66*, 1229-1238.
[39] Belingardi, G.; Cavatorta, M.P.; Paolino, D.S. (2006). Repeated Impact Behaviour and Damage Progression of Glass Reinforced Plastics.

Fracture of Nano and Engineering Materials and Structures. Proceedings of the 16th European Conference of Fracture, Alexandroupolis, Greece, July 3-7, 2006. Proceedings in CD-ROM.
[40] Sinmazçelik, T;. Arici, A.A.; Günay, V. *Jrnl Mater Sci.* 2006, *41*, 6237-6244.
[41] Yang, N.; Nayeb-Hashemi, H. *Jrnl Comp Mater.* 2007, *41*, 559-574.
[42] Azouaoui, K.; Ouali, N.; Ouroua, Y.; Mesbah, A.; Boukharouba, T. *Jrnl Sound Vibr.* 2007, *308*, 504-513.
[43] Casas-Rodriguez, J.P.; Ashcroft, I.A.; Silberschmidt, V.V. *Comp Sci Tech.* 2008, *68*, 2401-2409.
[44] Sinmazçelik, T.;. Çoban, O.; Bora, M.Ö.; Günay, V.; Cürgül, İ. *Appl Comp Mater.* 2008, *15*, 99-113.
[45] Belingardi, G.; Cavatorta, M.P.; Paolino, D.S. *Comp Sci Tech.* 2008, *68*, 2646-2652.
[46] Casas-Rodriguez, J.P.; Ashcroft, I.A.; Silberschmidt, V.V. *Comp Sci Tech.* 2008, *68*, 2663-2670.
[47] Casas-Rodriguez, J.P.; Ashcroft, I.A.; Silberschmidt, V.V. *Int Jrnl Sound Vibr.* 2007, *208*, 467-478
[48] Bora, M. Ö.; Çoban, O.; Sinmazçelik, T.; Cürgül, İ.; Günay, V. *Mater Des.* 2009, *30*, 145-153
[49] Ray, D.; Sarkar, B.K.; Bose, N.R. *Comp A* 2002, *33*, 233-241.
[50] Ray, D.; Sarkar, B.K.; Rana, A.K. *Jrnl Appl Polym Sci.* 2002, *85*, 2588-2593.
[51] Sarkar, B.K.; Ray, D. *Comp Sci Tech.* 2004, *64*, 2213-2219.
[52] Dhakal, H.N.; Zhang, Z.Y.; Richardson, M.O.W.; Errajhi, O.A.Z. *Comp Struct.* 2007, *81*, 559-567.
[53] Attia, O.; Kinloch, A.J.; Matthews, F.L. *Comp Sci Tech.* 2001, *61*, 2273-2283.
[54] Gassan, J.; Dietz, T. *Comp Sci Tech.* 2001, *61*, 157-163.
[55] Vassilopoulos, A.P.; Georgopoulos, E.F.; Dionysopoulos, V. *Int Jrnl Fat.* 2007, *29*, 20-29.
[56] Andersons, J.; König, M. *Comp Sci Tech.* 2007, *64*, 2139-2152.
[57] Ramkumar, R. L. In *Long - Term Behaviour of Composites;* O'Brien, T.K.; Ed.; ASTM STP 813; American Society for Testing and Materials: Philadelphia, PA, 1983; pp 116–135.
[58] Ramkumar, R. L. In *Damage in Composite Materials;* Reifsnider, K.L.; Ed.; ASTM STP 775; American Society for Testing and Materials: Philadelphia, PA, 1982; pp 184–210.
[59] Jurf, R.A.; Pipes, R.B. *Jrnl Comp Mater.* 1982, *16*, 386-394.

[60] Talbot, M.F.; Springer, G.S.; Berglund, L.A. *Jrnl Comp Mater.* 1987, *21*, 1056-1081.
[61] Hashemi, S.; Kinloch, A.J.; Williams, J.G. *Jrnl Comp Mater.* 1990, *24*, 918-956.
[62] Whitcomb, J.D. *Jrnl Comp Mater.* 1992, *26*, 1844-1858.
[63] O'Brien, T.K. In *Damage in Composite Materials;* Reifsnider, K.L.; Ed.; ASTM STP 775; American Society for Testing and Materials: Philadelphia, PA, 1982; pp 140-167.
[64] Wilkins, D.J.; Eisenman, J.R.; Camin, R.A.; Margolis, W.S.; Benson, R.A. In *Damage in Composite Materials;* Reifsnider, K.L.; Ed.; ASTM STP 775; American Society for Testing and Materials: Philadelphia, PA, 1982; pp 168-183.
[65] Threthewey, B.R.Jr; Gillespie, J.W.Jr.; Carlsson, L.A. *Jrnl Comp Mater.* 1988, *22*, 459-483.
[66] Beheshty, M.H.; Harris, B.; Adam, T. *Comp A* 1999, *30*, 971-987.
[67] Shen, F.; Lee, K.H.; Tay, T.E. *Comp Sci Tech.* 2001, *61*, 1239-1251.
[68] Turon, A.; Camanho, P.P.; Costa, J.; Dávila, C.G. *Mech Mater.* 2006, *38*, 1072-1089.
[69] Kim, S.J.; Hwang, I.H. *Int Jrnl Fat.* 2006, *28*, 1334-1339.
[70] Chen, A.S.; Almond, D.P.; Harris, B. *Meas Sci Tech.* 2001, *12*, 151-156.
[71] Chen, A.S.; Almond, D.P.; Harris, B. *Comp Sci Tech.* 2001, *61*, 2437-2443.
[72] Chen, A.S.; Almond, D.P.; Harris, B. *Int Jrnl Fat.* 2002, *24*, 257-261.
[73] Grondel, S.; Assaad, J.; Delebarre, C.; Moulin, E. *Ultrasonics* 2004, *42*, 819-824.
[74] Toyama, N.; Takatsubo, J. *Comp Sci Tech.* 2004, *64*, 1293-1300.
[75] Monnier, T. *Jrnl Intel Mater Syst Struct.* 2006, *17*, 411-421.
[76] Cesari, F.; Dal Re, V.; Minak, G.; Zucchelli,A. *Comp A* 2007, *38*, 1163-1173.
[77] Wang, S.; Chung, D.D.L.; Chung, J.H. *Jrnl Mater Sci.* 2005, *40*, 6463-6472.
[78] Wang, S.; Chung, D.D.L.; Chung, J.H. *Comp A* 2005, *36*, 1707-1715.
[79] Wang, S.; Wang, D.; Chung, D.D.L.; Chung, J.H. *Jrnl Mater Sci.* 2006, *41*, 2281-2289.
[80] Wang, S.; Chung, D.D.L.; Chung, J.H. *Jrnl Intel Mater Syst Struct.* 2006, *17*, 57-62.
[81] Wang, D.; Wang, S.; Chung, D.D.L.; Chung, J.H. *Jrnl Intel Mater Syst Struct.* 2006, *17*, 853-861.
[82] Wang, D.; Chung, D.D.L. *Smart Mater Struct.* 2006, *15*, 1332-1344.

[83] Wang, S.; Chung, D.D.L. *Carbon* 2006, *44*, 2739-2751.

[84] 84 Shen, L.; Li, J.; Liaw, B. M.; Delale, F.; Chung, J.H. *Comp Sci Tech.* 2007, *67*, 2513-2520.

[85] Todoroki, A.; Tanaka, M.; Shimamura, Y. *Comp Sci Tech.* 2005, *65*, 37-46

[86] Matsuzaki, R.; Todoroki, A. *Comp Sci Tech.* 2006, *66*, 407-416

[87] Ueda, M.; Todoroki, A. *Eng Fract Mech.* 2008, *75*, 2737-2750

[88] Angelidis, N.; Khemiri, N.; Irving, P.E. *Smart Mater Struct.* 2005, *14*, 147-154

[89] Angelidis, N.; Irving, P.E. *Comp Sci Tech.* 2007, *67*, 594-604

[90] Kuang, K.S.C.; Cantwell, W.J. *Jrnl Mater Sci Lett.* 2002, *21*, 1351-1354

[91] Kuang, K.S.C.; Cantwell, W.J. *Jrnl Thermopl Comp Mater.* 2003, *16*, 213-229

[92] Kister, G.; Ralph, B.; Fernando, G.F. *Smart Mater Struct.* 2004, *13*, 1166-1175

[93] Takeda, S.; Minakuchi, S.; Okabe, W.; Takeda, N. *Comp A* 2005, *36*, 903-908

[94] Silva, J.M.A.; Devezas, T.C.; Silva, A.P.; Ferreira, J.A.M. *Jrnl Comp Mater.* 2005, *39*, 1261-1281

[95] Chambers, A.R.; Mowlem, M.C.; Dokos, L. *Comp Sci Tech.* 2007, *67*, 1235-1242

[96] Takeda, S.; Aoki, Y.; Ishikawa, T.; Takeda, N.; Kikukawa, H. *Comp Struct.* 2007, *79*, 133-139.

[97] Pearson, J.D.; Zikry, M.A.; Prabhugoud, M.; Peters, K. *Jrnl Comp Mater.* 2007, *41*, 2759-2783.

[98] Sohn, M.S.; Hu, X.Z.; Kim, J.K.; Walker, L. *Comp B* 2000, *31*, 681-691.

[99] Guillaumat, L.; Batsale, J.C.; Mourand, D. *Comp A* 2004, *35*, 939-944.

[100] Mian, A.; Newaz, G.; Han, X.; Mahmood, T.; Saha, C. *Comp Sci Tech.* 2004, *64*, 1115-1122.

[101] Mian, A.; Han, X.; Islam, S.; Newaz, G. *Comp Sci Tech.* 2004, *64*, 657-666.

[102] Meola, C.; Carlomagno, G.M.; Sqillace, A.; Vitiello, A. *Eng Fail Anal.* 2006, *13*, 380-388.

[103] Wosu, S.N.; Hoy, D. *Jrnl Comp Mater.* 2006, *40*, 1577-1602.

[104] Carpino, G.; Lopresto, V.; Santoro, D. *Comp Sci Tech.* 2007, *67*, 325-335.

[105] Nayeb-Hashemi, H.; Cohen, M.D.; Zotos, J.; Poormand, R. *Jrnl Non-Destr Eval.* 1986, *5*, 119-131.

[106] Adden, S.; Pfleiderer, K.; Solodov, I.; Horst, P.; Busse, G. *Comp Sci Tech.* 2008, *68*, 1616-1623.
[107] Dunkers, J.P.; Sanders, D.P.; Hunston, D.L.; Everett, M.J.; Green, W.H. *The Jrnl Adh.* 2002, *78*, 129-154.
[108] Minnaar, K.; Zhou, M. *Jrnl Mech Phys Sol.* 2004, *52*, 2771-2799.
[109] Růžek, R.; Lohonka, R.; Jironč, J. *NDT&E Inter.* 2006, *39*, 132-142.
[110] Amaro, A.M.; de Moura, M.F.S.F.; Reis, P.N.B. (2006). Detection of Load Velocity Impact Damage in Carbon-Epoxy Plates using NDT. Fracture of Nano and Engineering Materials and Structures. Proceedings of the 16th European Conference of Fracture, Alexandroupolis, Greece, July 3-7, 2006. Proceedings in CD-ROM.
[111] Corigliano, A.; Ghisi, A.; Mariani, S. (2006). Detection of Load Velocity Impact Damage in Carbon-Epoxy Plates using NDT. Fracture of Nano and Engineering Materials and Structures. Proceedings of the 16th European Conference of Fracture, Alexandroupolis, Greece, July 3-7, 2006. Proceedings in CD-ROM.
[112] Hatsukade, Y.; Inaba, T.; Kasai, N.; Maruno, Y.; Ishiyama, A.; Tanaka, S. *Physica C* 2004, *412-414*, 1484-1490.
[113] Harris, B.; Chen, A.S.; Coleman, S.L.; Moore, R.J. *Jrnl Mater Sci.* 1991, *26*, 307-320
[114] Ma, C.-C.M.; Tai, N.H.; Wu, S.H.; Lin, S.H.; Wu, J.F.; Ong, C.L.; Wu, I.C.; Sheu, M.F. *Jrnl Pol Res.* 1994, *1*, 17-24
[115] Tai, N.-H.; Yip, M.-C.; Tseng, C.-M. *Comp B* 1999, *30*, 849-865
[116] Tai, N.H.; Ma, C.C.M.; Lin, J.M.; Wu, G.Y. *Comp Sci Tech.* 1999, *59*, 1753-1762
[117] Ding, Y.Q.; Yan, Y.; McIlhagger, R. *Jrnl Mater Proc Tech.* 1995, *55*, 58-62
[118] Mahinfalah, M.; Skordahl, R.A. *Comp Struct.* 1998, *42*, 101-106.
[119] Chotard, T.J.; Pasquiet, J.; Benzeggagh, M.L. *Comp Sci Tech.* 2000, *60*, 895-912.
[120] Margueres, Ph.; Meraghni, F.; Benzeggagh, M.L. *Comp A* 2000, *31*, 151-163.
[121] Chotard, T.J.; Pasquiet, J.; Benzeggagh, M.L. *Comp Struct.* 2001, *53*, 317-331.
[122] Symons, D.D.; Davis, G. *Comp Sci Tech.* 2000, *60*, 379-389.
[123] Horn, G.P.; Mackin, T.J.; Kurath, P. *Polym Comp.* 2001, *22*, 420-431.
[124] Melin, L.G.; Schön, J. *Comp Sci Tech.* 2001, *61*, 1841-1852.
[125] Melin, L.G.; Schön, J.; Nyman, T. *Int Jrnl Fat.* 2002, *24*, 263-272.
[126] Puhui, C.; Zhen, S.; Jun Yang, W. *Jrnl Comp Mater.* 2002, *36*, 589-610.

[127] Lee, S.-H.; Aono, Y.; Noguchi, H.; Cheong, S.-K. *Jrnl Comp Mater.* 2004, *38*, 1461-1477.

[128] Freeman, B.; Schwingler, E.; Mahinfalah, M.; Kellogg, K. *Comp Struct.* 2005, *70*, 374-381.

[129] Sanchez-Saez, S.; Barbero, E.; Zaera, R.; Navarro, C. *Comp Sci Tech.* 2005, *65*, 1911-1919.

[130] Petit, S.; Bouvet, C.; Bergerot, A.; Barrau, J.-J. *Comp Sci Tech.* 2007, *67*, 3286-3299.

[131] Xiong, J.-J.; Li, H.-Y.; Zeng, B.-Y. *Comp Struct.* 2008, *85*, 29-42.

[132] Yuanjian, T.; Isaak, D.H. *Comp B* 2008, *39*, 505-512.

[133] Saito, H.; Kimpara, I. *Comp Sci Tech.* In Press.

[134] Minak, G.; Morelli, P.; Zucchelli, A. *Comp Sci Tech.* In Press.

[135] Santulli, C.; Cantwell, W.J. *Jrnl Mater Sci Lett.* 2001, *20*, 477-479.

[136] Qi, B.; Herszberg, I. *Comp Struct.* 1999, *47*, 483-490.

[137] Nilsson, K.-F.; Asp, L.E.; Sjögren, A. *Int Jrnl Sol Struct.* 2001, *38*, 8407-8440.

[138] Olsson, R. *Comp Struct.* 2003, *59*, 341-349.

[139] Thomson, R.S.; Scott, M.L. *Comput Mech.* 2000, *26*, 75-89.

[140] Tay, T.E.; Shen, F. *Jrnl Comp Mater.* 2002, *36*, 1299-1320.

[141] Riccio, A.; Scaramuzzino, F.; Perugini, P. *Comp B* 2001, *32*, 209-218.

[142] Riccio, A.; Gigliotti, M. *Jrnl Comp Mater.* 2007, *41*, 1939-1960.

[143] Riccio, A.; Pietropaoli, E. *Jrnl Comp Mater.* 2008, *42*, 1309-1335.

[144] Kim, B.W.; Mayer, A. H. *Comp Sci Tech.* 2003, *63*, 695-713.

[145] Attia, O.; Kinloch, A. J.; Matthews, F.L. *Comp Sci Tech.* 2003, *63*, 1463-1472.

[146] Kang, K.-W.; Kim, J.-K. *Comp A* 2004, *35*, 529-535.

[147] Kang, K.W.; Kim, J.-K. *Comp A* 2006, *37*, 1451-1457.

[148] O'Brien, T.K.; Johnston, N.J.; Morris, D.H.; Simonds, R.A. *SAMPE Jrnl* 1982, 8-15.

[149] O'Brien, T.K. In *Effects of Defects in Composite Materials;* ASTM STP 836; American Society for Testing and Materials: Philadelphia, PA, 1984; pp 125-142.

[150] O'Brien, T.K.; Raju, I.S.; Garber, D.P. *Jrnl Comp Tech Res.* 1986, *8*, 37-47.

[151] Kuboki, T.; Gallagher, E.; Jar, P.-Y.B.; Cheng, J.J.R. *Appl Comp Mater.* 2005, *12*, 93-108.

[152] Andrews, M.G.; Massabò, R. (2006). Static and Dynamic Response of Multiply Delaminated Structures. Fracture of Nano and Engineering Materials and Structures. Proceedings of the 16th European Conference

of Fracture, Alexandroupolis, Greece, July 3-7, 2006. Proceedings in CD-ROM.
[153] Andrews, M.G.; Massabò, R.; Cox, B.N. *Int Jrnl Sol Struct.* 2006, *43*, 855-886.
[154] Icardi, U. *Int Jrnl Imp Eng.* 2007, *34*, 1835-1868.
[155] Watkins, S.E.; Akhavan, F.; Dua, R.; Chandrashekhara, K.; Wunsch, D.C. *Smart Mater Struct.* 2007, *16*, 515-524.
[156] Akhavan, F.; Watkins, S.E.; Chandrashekhara, K. *Mech Comp Mater Struct.* 2000, *7*, 195-205.
[157] Butler, R.; Almond, D.P.; Hunt, G.W.; Hu, B.; Gathercole, N. *Comp A* 2007, *38*, 1211-1215.
[158] Li, X.; Hallet, S.R.; Wisnom, M.R. *Comp A* 2008, *39*, 218-230.
[159] Guedes, R.M.; de Moura, M.F.S.F.; Ferreira, F.J. *Comp Struct.* 2008, *84*, 362-368.
[160] Craven, R.; Sztefek, P.; Olsson, R. *Comp Sci Tech.* 2008, *68*, 2518-2525.
[161] Bolotin, V.V. *Comp B* 1996, *27B*, 129-145.
[162] Bolotin, V.V. *Mech Comp Mater.* 2001, *37*, 367-380.
[163] Olsson, R.; Iwarsson, J.; Melin, L. G.; Sjögren, A.; Solti, J. *Comp Sci Tech.* 2003, *63*, 199-209
[164] Aoki, Y.; Ishikawa, T.; Takeda, S.i.; Hayakawa, Y.; Harada, A.; Kikukawa, H. *Int Jrnl Fat.* 2006, *28*, 1109-1115.
[165] Bennati, S.; Valvo, P.S. *Comp Sci Tech.* 2006, *66*, 248-254.
[166] Yuanjian, T.; Isaak, D.H. *Comp Sci Tech.* 2007, *67*, 3300-3307.
[167] Choi, S.W.; Song, E.-J.; Hahn, H.T. *Comp Sci Tech.* 2003, *63*, 661-675.
[168] Garg, A.K.; Roy Mahapatra, D.; Suresh, S.; Gopalakrishnan, S.; Omkar, S.N. *Comp Sci Tech.* 2004, *64*, 2477-2493.
[169] Wosu, S.N.; Hui, D.; Dutta, P.K. *Eng Fract Mech.* 2005, *72*, 1531-1558.
[170] Kim, J.-H.; Pierron, F.; Wisnom, M.R.; Syed-Muhamad, K. *Comp A* 2007, *38*, 2065-2075.
[171] Rhead, A.T.; Butker, R.; Hunt, G.W. *Int Jrnl Sol Struct.* 2008, *45*, 4349-4361.
[172] Mouritz, A.P. *Comp A* 2007, *38*, 2383-2397.
[173] Partridge, I.K.; Cartié, D.D.R. *Comp A* 2005, *36*, 55-64.
[174] Rugg, K.L.; Cox, B.N.; Massabò R. *Comp A* 2002, *33*, 177-190.
[175] Byrd, L.W.; Birman, V. *Comp B* 2006, *37*, 365-378.
[176] Dantuluri, V.; Maiti, S.; Geubelle, P.H.; Patel, R.; Kilic, H. *Comp Sci Tech* 2007, *67*, 616-631.
[177] Chang, P.; Mouritz, A.P.; Cox, B.N. *Comp A* 2007, *38*, 244-251.

[178] Chang, P.; Mouritz, A.P.; Cox, B.N. *Comp A* In Press, doi:10.1016/j.compositesa.2005.11.013.
[179] Mouritz, A.P. *Comp Sci Tech.* 2007, *67*, 3110-3120.
[180] Mines, R.A.W.; Roach, A.M.; Jones, N. *Int Jrnl Imp Eng.* 1999, *22*, 561-588.
[181] Aymerich, F.; Priolo, P.; San, C.T. *Comp Sci Tech.* 2003, *63*, 907-917.
[182] Aymerich, F.; Pani, C.; Priolo, P. *Comp A* 2007, *38*, 1174-1182.
[183] Hosur, M.V.; Karim, M.R.; Jeelani, S. *Comp Struct.* 2003, *61*, 89-102.
[184] Saito, H.; Kimpara, I. *Comp A* 2006, *37*, 2226-2235.
[185] Lopresto, V.; Melito, V.; Leone, C.; Carpino, G. *Comp Sci Tech.* 2006, *66*, 206-214.
[186] Velmurugan, R.; Solaimurugan, S. *Comp Sci Tech.* 2007, *67*, 61-69.
[187] Beier, U.; Fischer, F.; Sandler, J.K.W.; Altstädt, V.; Weimer, C.; Buchs, W. *Comp A* 2007, *38*, 1655-1663.
[188] Mouritz, A.P. *Comp Sci Tech.* 2008, *68*, 2503-2510.
[189] Tan, Y.; Wu, G.; Suh, S.S.; Yang, J.M.; Hahn, H.T. *Int Jrnl Fat.* 2008, *30*, 483-492.
[190] Kishore; Khan, B.; Viswanath, B. *Jrnl App Pol Sci.* 2002, *85*, 752-761.
[191] Zubeldia, A.; Larrañaga, M.; Remiro, P.; Mondragon, I. *Jrnl Polym Sci B Polym Phys.* 2004, *42*, 3920-3933.
[192] Lee, S.-H.; Lee, J.-H.; Cheong, S.-K.; Noguchi, H. *Jrnl Mater Proc Tech.* 2008, *207*, 21-29.
[193] Pegoretti, A.; Cristelli, I.; Migliaresi, C. *Comp Sci Tech.* 2008, *68*, 2653-2662.
[194] Gojny, F.H.; Wichmann, M.H.N.; Fiedler, B.; Schulte, K. *Comp Sci Tech.* 2005, *65*, 2300-2313.
[195] Sadeghian, R.; Gangireddy, S.; Minaie, B.; Hsiao, K.-T. *Comp A* 2006, *37*, 1787-1795.
[196] Wetzel, B.; Rosso, P.; Haupert, F.; Friedrich, K. *Eng Fract Mech.* 2006, *73*, 2375-2398.
[197] Arai, M.; Noro, Y.; Sugimoto, K.-i.; Endo, M. *Comp Sci Tech.* 2008, *68*, 516-525.
[198] Gassan, J.; Bledzki, A.K. *Appl Comp Mater.* 2000, *7*, 373-385.
[199] Sihn, S.; Kim, R.Y.; Kawabe, K.; Tsai, S.W. *Comp Sci Tech.* 2007, *67*, 996-1008.
[200] Lanciotii, A.; Lazzeri, L.; Raggi, M. *Comp Struct.* 1995, *33*, 87-94.
[201] Byrd, L.W.; Birman, V. *Int Jrnl Mech Sci.* 2006, *48*, 938-949.
[202] Kensche, C.W. *Int Jrnl Fat.* 2006, *28*, 1363-1374.
[203] Vogelesang, L.B.; Vlot, A. *Jrnl Mater Proc Tech.* 2000, *103*, 1-5.

[204] Alderliesten, R.C.; Hagenbeek, M.; Homan, J.J.; Hooijmeijer, P.A.; de Vries, T.J.; Vermeeren, C.A.J.R. *Appl Comp Mater.* 2003, *10*, 223-242.
[205] Vermeeren, C.A.J.R.; Beumler, Th.; de Kanter,J.L.C.G.; van der Jagt, O.C.; Out, B.C.L. *Appl Comp Mater.* 2003, *10*, 257-276.

INDEX

C

D

H

I

J

K

L

M

N

O

P

Q

R

S

T

U

V

W

Y